Managing Project Competence

The Lemon and the Loop

Best Practices and Advances in Program Management Series

Series Editor
Ginger Levin

Managing Project Competence

The Lemon and the Loop

Rolf Medina

CRC Press
Taylor & Francis Group
Boca Raton London New York

CRC Press is an imprint of the
Taylor & Francis Group, an **informa** business

AN AUERBACH BOOK

CRC Press
Taylor & Francis Group
6000 Broken Sound Parkway NW, Suite 300
Boca Raton, FL 33487-2742

First issued in paperback 2019

© 2018 by Taylor & Francis Group, LLC
CRC Press is an imprint of Taylor & Francis Group, an Informa business

No claim to original U.S. Government works

ISBN-13: 978-1-4987-8438-2 (hbk)
ISBN-13: 978-0-367-89021-6 (pbk)

Visit the Taylor & Francis Web site at
http://www.taylorandfrancis.com

and the CRC Press Web site at
http://www.crcpress.com

Dedication

I want to dedicate this book to everyone seeing a value in sharing knowledge, information, thoughts, ideas, secrets, and all other information that makes it easier for others to do their jobs.

Transparency makes the world a better place.

Table of Contents

List of Tables and Figures

Tables

Figures

Preface

A person who never made a mistake never tried anything new.

—Albert Einstein

Competence is a word that is used on almost a daily basis. We talk about who has the right competence, what competences we need to perform our work, that we lack certain competences, and much more. Most of us have a picture of what competence means; this picture might vary from time to time, and different persons might have a different picture. In 2005, Le Deist and Winterton stated that it is impossible to identify, or to find a coherent theory or definition capable of reconciling, all the different ways that the term *competence* is used. This book aims to sort out different dimensions of competence, outline some factors that make us competent, discuss practical aspects of how to develop competence, and give some examples of how an organization can utilize its competence in line with its strategy and goals. In this context, an organization can be a company, a department, or any entity with an organizational strategy and goals.

This book also takes the standpoint that projects are competence arenas. Projects are by their nature problem-solving activities, in which people develop their competences through "learning-by-doing." A project should have a distinct goal, moving something from point A to point B, and in this move, facing whatever challenges and solving whatever problems arise.

Projects can be an organization's proving ground for developing competence in line with the organization's strategies and goals.

Why is it important to write a book about competence today?

The answer to that question could naturally be that there is no coherent theory, as Le Deist and Winterton (2005) stated. But a more important answer is that when everything is moving faster, we have to be able to learn new things faster and to absorb new knowledge and skills continuously; in other words, we need to generate new competences faster and faster. To achieve continuous competence development, we need to both understand that competence has different dimensions and identify the factors that generate new competence.

Today, not many people learn a profession and stay in that profession for the rest of their lives. Instead, people move forward to new positions, taking on new roles and changing employers. The firms of today also act in a fast-moving market with new competition, acquisitions, global actors, longer value chains, new innovations in products, and new ways of working. Because of these facts, companies of today need to improve the utilization of their organizational competence in order to be innovative and competitive.

For companies to be successful, the management of an organization needs to understand how competence evolves and how it can be utilized and linked to the organization's goals. When the management understands this, there is a higher probability that the people working in the organization will be more satisfied with their working situation than when the management does not understand. The reason that satisfaction will increase is that competence will likely be central in the organization, with focus on motivating people to develop new competence, healthy internal mobility, and organizational learning. A good way of managing competence will in most cases lead to a win–win situation for the company and the employee.

Another perspective is that organizations tend to be more knowledge intensive than before, meaning that knowledge has a higher importance than other inputs and is based on the application of human capital and on the expansion of knowledge through problem solving, experimentation, or learning. People working in knowledge-intensive organizations are called *knowledge workers*— that is, people who work with knowledge as a base and have a need to continually renew their knowledge and competence.

Knowledge-intensive organizations tend to be project intensive, having many projects of different size and importance. These kinds of organizations use projects as an enabler for change and for developing new products or services. In this context, competence evolves through projects. Competitive advantage in these kinds of organizations depends on how well the organization can manage its knowledge and skills and create the ability to align them with the organization's strategic goals.

This book is based on the author's long experience in managing competence in organizations of different sizes and in different industries. This book is also based on empirical data from a multiple case study that ended up in a doctoral thesis. This study, containing a total of nine cases, was performed in three Swedish companies acting in different industries. The common attribute for the cases was that they were knowledge intensive to some extent, meaning that people were working in an environment in which they needed to apply knowledge and human skills.

To gain different perspectives on the topic, three types of organizations were studied. The first was a public sector organization; the second was a younger, private, fast-growing company in the technical product development field acting in a global market; and the third was a declining private company in the consumer electronics industry, also acting in a global market.

In addition, this book is based on an extensive review of literature in different domains, such as dynamic capabilities, learning, human resource management, project management, knowledge-intensive organization, and competence management. The combination of all these sources forms the content and the conclusions in the book.

Readers of this book are those who consider themselves to be knowledge workers. It will provide insights into the different dimensions of competence and the factors that generate new competence. Furthermore, it will present new views on how we can build competence development and agile performance management into daily work. Moreover, this book will support organizational leaders, functional managers, HR, project managers, and all people working in projects or other knowledge-creating activities. In addition, it is suitable for teaching and training purposes both for universities and for professionals. Practical examples combined with different methods and approaches will guide putting theory into practice. This book also presents new approaches, frameworks, and methodologies, such as the competence lemon, the competence loop, and REPI.

If you want to understand how you can develop your competence and how an organization can be effective in managing competence and being innovative, this book is for you.

Acknowledgments

Writing a book about competence means that you constantly are exposed to the subject. People are more than willing to discuss what competence is and what skills and personal characteristics are. I am grateful to all of you who have given me so much input and reflections. I include colleagues, customers, suppliers, family, friends, and many more.

This book is based on my own experience, on deep and extensive review of the literature, and on multiple case studies. I want to thank my primary advisor, Dr. Ginger Levin, for her outstanding support and engagement. Her guidance and feedback encouraged me and helped me go in the right direction. I am also very grateful to Dr. Alicia Medina for being my wife and part of my supportive family, as well as an excellent secondary advisor. Her feedback and support has been invaluable and cannot be expressed in words.

The multiple case studies would not have been possible without great help from the studied organizations. Anna, Ulf, Mats, and Patrick, you know how grateful I am for opening the doors and connecting me with the right people. And then we have the man with the network, Jan Raaschou, who has help me with many contacts and the focus groups.

I also want to express my appreciation to my editor, John Wyzalek, for his excellent guidance in making this book possible. In addition, I want to acknowledge the production team and especially Susan Culligan, for very good cooperation during the production of the book.

My family is always my indispensable fundament for everything I do and gives me energy, positive feelings, and creativity.

To all, my sincere thanks.

About the Author

Rolf Medina in an independent management consultant and coach who has worked with many large enterprises, such as IKEA, Ericsson, and Sony, with organizational development, strategic directions, managing different kind of change projects, and—last but not least—with establishing new ways of managing competence in which there is a win–win situation between the co-workers and the company. Besides consultancy and coaching assignments, Rolf has been manager at different organizational levels in both small firms and larger international corporations, always in a knowledge-intensive environment.

In addition to his professional career, Rolf has conducted research in organizational development and innovation, in which competence and projects have been cornerstones leading to several academic publications in well-ranked journals. His previous research is one of the bases for the present book.

In parallel with his professional career, Rolf is guest lecturer and supervisor at Umeå University in Sweden, SKEMA Business School in France, and University College of London in the UK.

Rolf holds a PhD in Programme and Project Management from SKEMA, France; a MSc in Project Management from Karlstad University in Sweden; and a BSc in Computer Engineering and Mechatronics from University College of Halmstad in Sweden.

Rolf is married to Dr. Alicia Medina, and the family also consists of five children—Claudio, Adrian, Josefin, Veronica, and Gabriela. In addition to these children, there is also a new little person, Leon, his grandchild. His family means everything to him, and he gets inspiration from the time they all spend together.

As reflected in the book, Rolf is passionate about people's growth and that we shall open up our minds to make a better world. This leads to a deep interest in nature and how we manage climate change. This interest in nature led, in 2015, to his and Alicia's walking "el camino de Santiago," more than 800 kilometers, in 30 days. Walking every day gives time to reflect, which is a key point in this book. A major part of the book evolved during the 800-kilometer walk, while reflecting on how can we manage competence in a way that contributes to both the organization and to the people themselves.

Chapter 1

The Competence Lemon – Different Dimensions of Competence

1.1 What Is Competence?

I have been impressed with the urgency of doing. Knowing is not enough;
we must apply. Being willing is not enough; we must do.
— Leonardo da Vinci

In many cases, we use the word *competence* when we mean *knowledge*—we refer to a person with specific knowledge and experience in a subject matter area that is useful to perform some kind of work as "competent." Competence is in many cases used interchangeably with competency (Teodorescu, 2006). Generally, competency refers to behavioral areas, whereas competence is related to functional areas, but the usage can be inconsistent (Le Deist and Winterton, 2005). In general, competency is a set of behaviors a person must have, and it has a *worker*-orientation perspective, whereas competence is needed to perform tasks required in a job and has a *work*-orientation perspective (Chen and Chang, 2010). But competency and competence are two sides of the same coin, and both words can be used synonymously.

1.1.1 Types of Knowledge

Knowledge and experience form the basis for competence; without them we can hardly be competent in any area. In 1958, Polyani coined the term *tacit knowledge* to distinguish knowledge that is embodied in practice from knowledge that can be encoded and stored—namely, *explicit knowledge.* Explicit knowledge is built into working processes, documentation, information, etc., whereas tacit knowledge is *implicit*—it is what we have in our subconscious minds. Many of us learned to ride a bicycle when we were young, and we have that knowledge in our minds, but we cannot really explain how we ride the bicycle. Nonaka (1994) further developed the concept of tacit and explicit knowledge by arguing that the different types of knowledge could be placed along a continuum. He also proposed four different modes of knowledge conversion that pertain to the interaction between tacit and explicit knowledge:

1. **Socialization,** wherein tacit knowledge is converted into other tacit knowledge based on interactions between individuals. Sharing of knowledge is, according to Nonaka (1994), extensively dependent on people's shared experience, even if it is difficult to share each other's thinking processes without having shared experience. This mode could be connected to organizational culture. Moreover, this knowledge conversion mode is applied in daily work when we have working meetings, discuss problems with colleagues, and share our findings with others. One person transfers his or her knowledge to another.
2. **Combination,** wherein individuals combine and exchange explicit knowledge through meetings, presentations, and other similar mechanisms. This way of exchanging knowledge involves documented knowledge that is shared in the meeting or through a presentation. The knowledge is documented, and the receiver can take it in by watching the presentation and/or by receiving some kind of documentation.
3. **Externalization,** wherein tacit knowledge is converted into explicit knowledge. Within a project, we acquire new knowledge that we codify by documenting the findings. Documentation of a product or service is a typical example of externalization.
4. **Internalization,** wherein explicit knowledge is transferred into tacit knowledge. This could be considered as the traditional view of learning by training or courses. In the training, the teacher shows presentations and the students have books and other literature from which they draw conclusions. In this way, they acquire new tacit knowledge based on explicit knowledge from the literature and from the presentations.

The four knowledge conversion modes could be seen as different learning activities during which an individual's knowledge base grows. We use socialization when we share ideas with each other, combination when we show a PowerPoint presentation in a meeting, externalization when we document lessons learned in a project, and, finally, internationalization when we attend a course.

Starbuck (1992) makes a similar distinction between *esoteric* and *common* knowledge. Expertise is based on esoteric knowledge, which in turn gives rise to power. When knowledge is less esoteric, its ability to give rise to power diminishes. Looking at Starbuck's viewpoint, the experts have power as long as they have unique knowledge that they have to maintain. From this perspective, specialization is preferable to spreading knowledge to the team level or to other organizational levels.

Interest in the power effect of knowledge has to some extent been overshadowed by the interest in the distinction between explicit and tacit knowledge (Kärreman, 2010). The reason behind Kärreman's argument is probably that Polyani's and Nonaka's explicit and tacit knowledge concepts are easier to take in and use in our daily life. Another reason is that it is important to renew our knowledge and work together with others. Specialization is for the few and knowledge sharing for the many!

1.1.2 Knowledge and Competence

Something more than knowledge and experience is needed to be able to use these in a way that adds value and makes us able to perform some kind of work. According to Wright, Dunford, and Snell (2001), competence is held by individuals and refers to work-related knowledge and skills and the ability to use them. Eden and Ackermann (2010) emphasize that a statement prefaced with the phrase "ability to" describes competence.

The ability to do something is linked to performance, which for Sanford (1989) means just the ability to apply knowledge and skills. Spencer, McClelland, and Kelner (1994) also emphasize that competence is based on knowledge and skills, but they add the attitudes required for performance in a designed role and setting. The latter view is closely related to that of Turner and Müller (2006), who argue that, aside from knowledge and skills, personal characteristics are also a part of competence.

In line with the above, the Project Management Institute (PMI, 2007) mentions the following major components as parts of competence: abilities, attitudes, behavior, knowledge, personality, and skills. It defines competence as "a cluster of related knowledge, attitudes, skills, and other personal characteristics that affect a major part of one's job" (PMI, 2007, p. 73). The framework also states that competence can be measured against predefined standards and improved by training and development. In addition, PMI distinguishes between *knowledge,* as knowing something, and *skills*—namely, the ability to use knowledge and a developed aptitude.

1.1.3 Input and Output Competence

PMI's definition of competence is in line with Crawford's (2005), in which competence is divided into three components: *input* competences, *personal* competences, and *output* competences. By input competences, Crawford means a person's knowledge and skills, whereas personal competences are core personality characteristics that a person needs to do a job. In Crawford's model, output competences are related to performance and the individual's ability to perform activities in relation to expected performance.

Eden and Ackermann (2010) emphasize that it is important to distinguish between competence and its outcomes, but it is easier to define the outcomes. A competence outcome cannot be managed directly, solely through the competences that create the outcome, although it is the competence outcome that supports a goal. Furthermore, a distinctive competence is considered as a manageable resource, whereas competence outcomes are results of the management of resources (Eden and Ackermann, 2010). In addition, Teodorescu (2006) emphasizes that competence itself cannot be measured; it is, rather, results and outcomes that have measurable attributes. Managers of companies are in general interested in the activities and behaviors that add value to their organizations, which are the results of activities—that is, the outcomes of competence (Gilbert and Cordey-Hayes, 1995; Teodorescu, 2006).

1.1.4 Knowledge- and Social-Based Competence

Koskinen (2015) has a similar approach that divides competence into knowledge- and social-based competence, wherein an individual's tacit and explicit knowledge form knowledge-based competence, and social-based competence consists of the ability to combine feeling, thinking, and acting in order to achieve results through social activities valued in the organizational context and culture. By using Koskinen's view, we extend the competence definition to include interaction with others, and not only what a person is able to do or perform.

But is competence an asset, or does competence exist only when a person is performing a task in a specific context?

1.1.5 A New View of Competence

The traditional view described above is based on the view of competence as an asset. However, Von Krogh and Roos (1996) brought forward a different view, emphasizing that competence means the intersection between a specific task and the knowledge and skills of the person or the team. Competence only exists when knowledge and skills are used and meet the task. A conclusion of this reasoning could be that the context and the task have an impact on competence independent of whether we see competence as an asset or an event. This is in line with Le Deist and Winterton (2005), who also emphasize that competences are centered on the individual, but that people do not have competences independent of the context. Also, Koskinen (2015) drew a connection between competence and context. Is a competent person in one context also competent in another? We will go further into this question in Chapter 6, in which the application of competence will be analyzed in different contexts.

Based on the above reasoning, it can be concluded that competence is based on knowledge, skills, personal characteristics, and social interactions, but that it is also related to a person's demonstrable performance, which can be measurable.

The limitation of, and maybe the problem with, this definition is that it gives a static view of competence. We have knowledge and skills and can apply them, which has a measurable outcome. But competence is also related to acquiring new knowledge—a dynamic and sustainable perspective on competence. Sustainable competence is needed to adapt to new conditions and new ways of working.

1.1.6 The Knowledge-Intensive Company

Looking at organizations in general, we can see that they are becoming more knowledge intensive, meaning that knowledge has more importance than other inputs and that human capital dominates (Starbuck, 1992). All organizations are to some extent built on knowledge, but a knowledge-intensive organization also tends to be ambiguity intensive in the sense that these kinds of organizations work with a higher degree of uncertainty (Alvesson, 2011). The knowledge-intensive economy is increasingly growing (Sinha and Van de Ven, 2005), and the successful companies will be the ones that manage their knowledge development and consider what knowledge means in their organizations (Von Krogh and Roos, 1996). Furthermore, Alvesson (2000) mentions different kinds of knowledge-intensive organizations, such as R&D, consultancy, etc., whereas organizations such as manufacturing firms are considered to be less knowledge intensive.

Based on this argument, it can be concluded that a company can have different levels of knowledge intensity in different parts of the organization, meaning that not all parts within a company need to have the same level of knowledge intensity; for instance, in a manufacturing firm, some parts of the organization, such as R&D, are more knowledge intensive than other parts, such as working at a production line.

As stated earlier, there needs to be a more dynamic view of competence. As previously concluded, human performance is related to the application of knowledge, which leads to some kind of outcome or result. Personal characteristics facilitate the application of knowledge. In addition, competence in a knowledge-intensive context not only means how a person can apply the knowledge and experience, but also how able they are to acquire new knowledge. The latter part changes the view of competence as something static to competence as something that constantly needs to be renewed, or maybe to something that facilitates renewal of knowledge. Renewal of knowledge lifts our view of competence to look at it from different perspectives and define it as *sustainable competence,* in which focus is on both application and renewal of knowledge.

Combining the traditional static view of competence with the dynamic and sustainable view, the next section will look at different dimensions of competence.

1.2 Six Dimensions of Competence

Nothing will work unless you do.

— Maya Angelou

In the last section, we concluded that competence is built on knowledge and experience, but also that there are other dimensions of competence facilitating application of knowledge and experience and the ability to acquire new knowledge, and in this way make competence sustainable.

Figure 1.1 Six dimensions of competence.

What are those other dimensions of competence?

Based on an extensive review of the literature combined with empirical data from case studies, six dimensions of competence emerged—namely, knowledge and experience, personal capability, social capability, leadership qualities, ability to learn, and ability to manage complexity, which can be seen in Figure 1.1. Before going into each of these dimensions, it is worth emphasizing that the dimensions differ depending on in which context the person acts, which will be further discussed in Chapter 3 and Chapter 6.

1.2.1 Knowledge and Experience

The first dimension of competence is *knowledge and experience,* which means that a person needs knowledge to be able to solve a problem or task in a specific subject matter area. For example, if he or she works as a purchaser, knowledge and experience could be knowledge in negotiation, purchase processes, legal, the specific product, etc. Knowledge in a subject matter area can be technical knowledge, business knowledge, process development, or some other area. If we do not use our knowledge, it tends to become obsolete over time (Cabello-Medina, López-Cabrales, and Valle Cabera, 2011). Knowledge and experience are mainly from the area in which we perform our work, but knowledge and experience from adjacent subject matter areas can enrich our competence when performing our work. If we have worked with logistics processes, this experience can improve our understanding when developing new manufacturing processes—we understand the adjacent processes because the two processes have many similarities and are related to each other.

1.2.2 Personal Capabilities

The second dimension of competence, *personal capabilities,* consists of two parts.

- The first part is an individual's personal characteristics, such as being pedagogical, innovative, effective, etc., which impacts how they use their knowledge and experience to solve problems. If a person works with technical problem solving, analytic skills will increase his or her ability to solve the problem.
- The second part of personal capabilities is the person's attitude toward work, such as being responsible in completing tasks, acting professional, doing what is expected or more, and being safe in what they are doing. A positive attitude toward work will increase motivation and the ability to do a good job.

Personal capabilities are the basis for understanding and solving a problem. A project manager summarized his view of personal capability as: "The ability to use theoretical knowledge in practical situations." An example is a teacher who needs to have pedagogical skills to perform teaching activities well, but will also be helped by having a positive attitude toward work and students, which will make the teacher more competent, since his or her performance will probably increase.

1.2.3 Social Capability

Having knowledge and experience and personal capabilities leads to having the ability to solve problems on one's own. In most cases we are interacting with other people in our daily work. *Social capability* is the third dimension of competence and is the ability to share knowledge and interact with others. This includes the ability to listen and be open to others' ideas and opinions, and also your own ability to explain your knowledge to others. High social capability increases the ability to cooperate with others in a productive way. Social capability also includes a person's skills in networking and knowing who knows what. Knowledge-intensive work needs, in many cases, to be performed in interaction with others—for example, in project teams solving problems together. Good social capability will increase successful problem solving in groups. A high level of social capability is needed when a person operates within a high level of interactions—for instance, a sales representative or a project manager for a project with many stakeholders. On the other hand, the system developer does not need to have the same level of social capability, because a large amount of the work needs to be done on their own.

Knowledge and experience, personal capabilities, and social capability all relate to application of knowledge to solve a task related to performance, which is the traditional view of competence. The remaining three dimensions are more focused on renewal of knowledge, although they also are important for the application of knowledge.

1.2.4 Leadership Qualities

In many cases, a person needs to provide information to others, enabling them to solve a problem, which requires the person in such situations to have *leadership qualities,* the fourth dimension of competence. Leadership qualities can be people management skills, having the capability to lead a team in the right direction, but also the ability to lift and support others in their work.

For a project manager, those qualities could mean the ability to manage different kinds of projects, suppliers, or heterogeneous teams. The project manager's leadership qualities may differ between different types of projects. If the project is a software development project, the leadership qualities focus on developing good team spirit, following up on the team members' deliveries, and communicating with relevant stakeholders. On the other hand, if the project consists of deliveries from various suppliers, the project manager's leadership qualities should lie in supplier management and stakeholder management.

Also, other people need to have leadership qualities. For example, a system architect needs to be able to support and coach system developers in their work, which requires the ability to lead, communicate, and share relevant information with others.

1.2.5 Ability to Learn

In today's business world, people are continuously facing new situations and problems. In those situations, they need to be good at learning new things, which is the fifth dimension of competence—namely, the *ability to learn*. This dimension of competence involves the capacity to take in and interpret new knowledge. Other aspects of this dimension of competence are knowing how to get information and being well informed in the subject matter area.

The nature of a project is to deal with new situations in terms of new products, changed ways of working, new markets, and other changed situations. New knowledge needs to be gained to fulfill project goals. Knowledge in this case does not only include knowledge of the specific subject matter area, but can also, for example, involve knowing how to manage stakeholders in a more efficient way or developing skills in presenting and selling the new solution. Having the ability to learn will increase a person's capacity to take in new knowledge and use it in a productive way. One example is a product designer who needs to learn new material and design styles to be able to meet the market demands of new products.

1.2.6 Ability to Manage Complexity

The last dimension of competence is the *ability to manage complexity*. This dimension of competence looks at the capability to manage ambiguity and complex, constantly changing situations, such as complex stakeholder situations, many different suppliers, etc. It also means the ability to take in information and link different domains to a conclusion.

People are receiving information from different sources and domains and need to combine those into one worldview. With the ability to manage complexity, they have a holistic view and see things from different angles, and they are capable of thinking several steps ahead and foreseeing consequences. Another aspect is being able to make decisions based on facts by balancing advantages and disadvantages. In many situations, it is necessary to make decisions or solve problems without having all information in place. One example of this situation is a vendor evaluation. Vendors can be evaluated based on many criteria, but it is impossible to gather all information about all the vendors. In some situations, there is a need to make a decision, otherwise the evaluation will keep on going. High ability to manage complexity will increase the ability to make that decision. Awareness of this dimension of competence will also create understanding of why people in the organization are reluctant to make decisions without having all information in place.

Another example of the ability to manage complexity is using knowledge from one domain to solve a problem in another. Managers can use knowledge from manufacturing processes to improve public sector administration. One example of the latter is using a lean approach in administrative work.

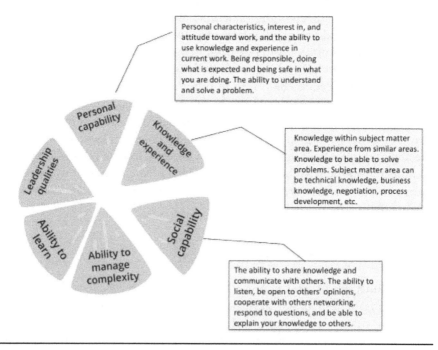

Figure 1.2   Summary of personal capability, knowledge and experience, and social capability.

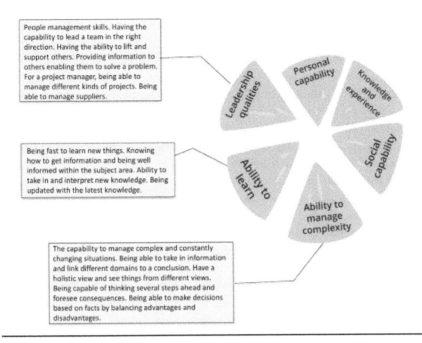

People management skills. Having the capability to lead a team in the right direction. Having the ability to lift and support others. Providing information to others enabling them to solve a problem. For a project manager, being able to manage different kinds of projects. Being able to manage suppliers.

Being fast to learn new things. Knowing how to get information and being well informed within the subject area. Ability to take in and interpret new knowledge. Being updated with the latest knowledge.

The capability to manage complex and constantly changing situations. Being able to take in information and link different domains to a conclusion. Have a holistic view and see things from different views. Being capable of thinking several steps ahead and foresee consequences. Being able to make decisions based on facts by balancing advantages and disadvantages.

Figure 1.3   Summary of leadership qualities, ability to learn, and ability to manage complexity.

1.2.7 Summary of the Six Dimensions

The six dimensions of competence are interrelated—for example, personal capabilities such as curiosity and attitude, as well as the ability to manage complexity, moderate the ability to learn. Figures 1.2 and 1.3 summarize the six dimensions of competence. With the six dimensions of competence in mind, the next section will look into some factors that have an impact on generating new competences. These factors are what makes competence dynamic and sustainable, and connects competence and learning.

1.3 Factors That Have an Impact on Generating New Competences

> *I've learned that people will forget what you said, people will forget what you did, but people will never forget how you made them feel.*
>
> — Maya Angelou

As concluded, competence is not static; instead, people continuously develop new competences involving different kinds of activities. Generating new competences is in fact learning, in which one acquires new knowledge and also develops other dimensions of competence. Learning as such can be seen both as a permanent change in behavior based on experience and reception, which leads to better performance (Holmqvist, 2003; Gunsel, Siachou, and Acar, 2011), and also as an organization's ability to acquire resources by acquisition of new resources or knowledge on how to use existing resources (De Wever, 2008).

Szulanski (2000) elaborates on the learning process by referring to the experimentation or planning stage before the actual use of knowledge as "learning before doing" (p. 12), and the phase during which knowledge is put to use, entailing the resolution of unexpected problems, as "learning by doing" (p. 12). Another view of learning is the "learning by working" approach, although learning can be seen as knowledge creation through social participation in everyday work—for example, in project teams transcending organizational boundaries (Fenwick, 2008). This is in line with Holmqvist (2003), who states that learning is a social activity in which individual learning takes place in social contexts.

Learning from an organizational perspective is a process (Oltra and Vivas-López, 2013) of acquiring, transferring, and integrating new knowledge and, in this way, adding value to the organization (Jerez-Gómez, Céspedes-Lorente, and Valle-Cabrera, 2005). Organizational learning can be carried out by both *informal* and *formal* processes; the informal processes occur when people share knowledge in daily work, and the formal processes are the means by which the organization integrates knowledge from individual to group to organizational level, and in this way either expanding existing knowledge or creating new knowledge (Lin, McDonough, Lin, and Lin, 2013).

To distinguish between individual and organizational learning, the focus in organizational learning is on the *knowledge* dimension of competence, whereas individual learning develops *all* dimensions of competence. The reason for this is that the dimensions besides knowledge are personal and a part of the individual's skill set, whereas knowledge is something that can be explicitly stored. If skills such as pedagogical aptitude are considered to be stored and shared in an organization, they can be considered as knowledge about how to be pedagogical.

One important contribution to defining factors that have an impact on generating new competence was a multiple-case study in three organizations that consisted of nine cases, which were mentioned in the introduction and will be further described in Chapter 3. The case studies showed that factors that enable learning and have an impact on generating new competence could be on either an organizational or a personal level. Factors on an organizational level are those that involve learning by interaction between individuals and are to a large extent also dependent on the organizational culture. On the other hand, personal factors are those that are closely connected to personal capabilities in the competence level and that increase learning. On an organizational level, factors that support generating new competence through interpersonal interactions are *sharing, social context, group learning,* and *heterogeneous environment,* whereas *attitude, problem solving, reflection, time to learn, taking responsibility,* and *training* are personal-level factors that enable people to generate new competences based on personal preferences and activities. These factors can be seen as the basis for learning in a professional context. They are summarized in Table 1.1 (on page 19).

1.3.1 Operational Factors

Sharing

The first factor on an organizational level is *sharing,* a competence-generating factor in which people share ideas and proposals, discuss solutions, and work together. As a result of sharing, both the one that shares information and the receiver(s) of the information can learn from the discussion. Another view of sharing occurs when the organization is transparent in terms of access to information, so that people can obtain information from other parts of the organization, and in this way learn by bringing in different perspectives.

Based on the six dimensions of competence described above, we can see that both personal and social capabilities are important for being good at sharing. A person having a positive attitude and personal characteristics supporting sharing knowledge with others will probably be better at sharing than a person lacking these characteristics. Having a high level of social capability will of course facilitate interaction with others.

The Social Context

The *social context* is an important factor when generating new competence. A context that stimulates work, in which people are seen and obtain feedback, has a positive impact on learning. An innovative environment inspires people to take responsibility and learn. In a learning environment, people feel trust in the team and in the organization, are allowed to make mistakes, and have the freedom to learn. Leaders support people in the learning process, and there is space to develop and work on new ideas. Positive and functional organizational cultures in which people feel trust in each other are the foundation for a positive social context.

In terms of the importance of competencies, personal capabilities are the most important to learn in different social contexts. A strong person can act in a negative social context, but a person with low self-confidence or who is shy needs to have a positive social context to dare to test new ways of solving problems or learn in different ways.

Group Learning

Another factor on an organizational level is *group learning,* which is an activity in which people work together and learn in-group by discussing solutions and how to reach the goal. A good team composition creates an environment of positive group learning in which people feel safe to share and discuss ideas. In such an environment, learning stimulates and the team shares ideas spontaneously.

For this factor, different dimensions of the competence mentioned above are important. The team leader needs to have good leadership qualities in terms of enabling the team to collaborate and share knowledge. As for sharing, a high degree of social capabilities facilitates sharing in groups. The same applies to personal capabilities; having attitudes and personal characteristics that support knowledge sharing and collaboration makes group learning more efficient. A student team working on their teamwork is a typical example of group learning taking place. Other examples are project teams working on solving problems together instead of working independently.

In a project context, Walker and Lloyd-Walker (2015) coined the term *co-working* to refer to situations in which people explain concepts and plans for each other. In this situation, there are two types of conversations.

- The first conversation is with themselves, preparing in their mind what to say and how to explain for the others. This conversation will be a self-reflective activity, filtering thoughts and ideas and forming a reality.
- The second conversation is one in which the listener takes in new information and reflects on it in relation to his or her own reality. In this way, dialog can be a very creative activity in which all involved parties learn from the dialog as such.

However, to get efficiency in a group learning activity, all participants need to respect each other and each other's knowledge. For group learning, the REPI methodology (Reflection, Elaboration, Participation, and Investigation) described in Chapter 4 is suitable.

Heterogeneous Environment

If different competences are brought together into a *heterogeneous environment,* learning can be even more effective. Cross-functional teams working with suppliers or customers make it possible for different perspectives to be analyzed and for the individual to be exposed to different perspectives on the topic, thereby learning a great deal. In such an environment, it is important to establish a culture in which people have an interest in the others' areas.

An example of a heterogeneous environment is a cross-functional project team with members from different departments. The team can consist of representatives from finance, legal, procurement, supply chain, IT, or other functional areas. The team members can be seen as experts in their individual subject matter areas. In many cases the team members do not fully respect the others' knowledge or expertise; from their perspective, the most important area is their own subject matter area. If the team leader can make the team members respect each other's subject matter areas and share their knowledge, learning will increase significantly, and teamwork will be more efficient.

Leadership qualities are an important competence dimension. The team leader needs to get the team together and get them to share knowledge and respect each other. A heterogeneous team has a higher level of complexity than a homogeneous team, because people have either different knowledge or different experiences.

High ability to manage complexity will facilitate learning in a heterogeneous environment; the person can combine knowledge from different areas into new knowledge. Also, the ability to take in information from people from different organizational entities will increase learning. As for group learning, if the team members have good social capabilities, learning will increase as the team members interact better with each other.

Time to Learn

In many cases, we lack time to reflect, to test new ways to solve a problem or take on an issue. Instead we take the easy way out and do as we used to do. The last factor on an organizational level is *time to learn*—time and space to seek information and learn from it. This factor is also impacted by the organizational culture. Do we have a culture in which time to learn is allowed or not?

We can see that in some organizations, time is important, whereas others put more importance on quality. In one of the companies studied, time is of high importance, meaning that there is high pressure to deliver as fast as possible, which has an impact on quality and on time to reflect and share information with others. The importance of time was expressed by a project team member as: "Time is important, we have to deliver in time. Some consequences are that competence development has low priority, and we lack time to share our experiences and knowledge."

In another company studied, quality is more important than time. This company encourages time to learn and innovate. Different departments have different activities to support innovation; most of them allocate time for innovation days or similar activities. For example, one of the departments has time set aside one week, twice a year, in which the employees are able to work with challenging new areas, experiment, and test new product ideas.

These kinds of activities encourage learning and the development of new competence. One project team member explained time to learn in a simple way as: "To learn, you need to have time and space to try new ways and search for information."

The impact of factors on an organizational level that facilitate learning confirms Holmqvist's (2003) statement that learning is a social activity in which individual learning takes place in social contexts. Fenwick (2008) also argues that learning can be seen as knowledge creation through social participation in everyday work—for example, in project teams that transcend organizational boundaries, which links social factors to project context.

The other set of factors that have an impact on generating new competence are on the personal level; they are also a part of the personal capabilities outlined among the six dimensions of competence mentioned above.

1.3.2 Personal Factors

Attitude

The first factor is *attitude;* people with a positive attitude have engagement and interest in the subject area, and a desire to do their best. With willingness and ambition to develop oneself, motivation is increased and new competence is generated. Taking responsibility, being curious, and having an interest in learning are also attributes in this factor. Attitude and motivation are closely connected to each other. With a negative attitude toward work, someone will not be motivated to do a good job. Attitude is a part of personal capabilities from the six dimensions of competence mention above. A person with a low level of knowledge but with a positive attitude toward work will learn more and be more efficient than a person with a higher level of knowledge but a negative attitude toward work. For that reason, having a positive attitude will increase competence generation.

Problem Solving

The second factor, *problem solving,* is activated when people are being exposed to, and trying to find solutions for, new and challenging working tasks and situations. When solving problems that one does not how to address in the beginning, new competence will be generated. The factor also relates to trial and error, testing different ways of doing things, and taking on new tasks in projects or other kinds of work situations. To solve problems, we have to acquire new knowledge.

An example is an engineer who has a technical challenge to tackle and needs to seek information about the challenge, maybe experiment with different solutions and test different scenarios. In this way, the engineer acquires new knowledge in the technical area and also learns new ways to tackle the problem and seek information.

Besides personal capabilities, the ability to manage complexity and the ability to learn are two important dimensions of competence mentioned above. Combining different kinds of information into new knowledge is a part of the ability to manage complexity. In the example, the engineer uses different information and tests different ways of solving the problem. A high level of ability to learn facilitates taking in new information, interpreting it, and generating new knowledge. In this example, the engineer creates and conducts tests, learns from the tests, and takes the next steps in solving the problem.

Reflection

Another factor on the personal level that has an impact on generating new competence is *reflection*. It involves people contemplating how tasks were solved, what decisions were made, and why the decisions were made. We also learn when we reflect on mistakes or experiences from performed work. Another aspect of reflection is when we prepare a presentation and reflect on how the audience will take in the information. By reflecting on issues and tasks, we also think about what to avoid and what to redo, and, in this way, we learn different ways of doing things. As described in the group learning factor above, preparing a presentation is often a self-reflective activity. We reflect on how we are going to present and also filter the information. The listener also reflects on received information and tries to put it in his or her own context and reality.

Besides personal capabilities, high levels of ability both to manage complexity and to learn make reflection more efficient from a learning perspective. While reflecting on what we have done, we draw conclusions by linking information from different areas into new knowledge, which is a part of the ability to manage complexity. We can also draw conclusions from different kinds of stakeholders and learn from them. Also, a high level of ability to learn increases learning where reflection can bring new insights and learnings. In the example above with the engineer who was solving problem, he or she conducts tests, reflects on the results, and learns from that.

The first letter of the REPI model described in Chapter 4 stands for *reflection* and is an important factor for learning.

Taking Responsibility

When getting a new job or taking on a new task, we are *taking responsibility* and learning from this new situation. In the new role or working with the new task, one needs new knowledge and maybe different kinds of personal skills, depending on the new situation.

An example is getting a new job in a new department or being appointed as project manager for a new and challenging project. In these situations, workers are exposed to new responsibilities and new work tasks, and in most cases are motivated and thrilled to learn new things. The new role should be a reasonable challenge to facilitate a high degree of learning. If the challenge is too big, the person will probably not learn that much, and their motivation will go down.

Personal capabilities will impact how much someone can learn by taking on new responsibilities. The other dimension of competence mentioned above that has a significant impact on generating new competence by taking responsibility is leadership qualities. If someone is appointed as project manager for an important project, their ability to understand the new project and the new team will increase their learning.

Training

The last factor on a personal level that has an impact on generating new competence is *training*, which is related to formal training and courses. In many cases, attending a course gives a person new ideas and perspectives on problems or daily work tasks. Applying the new ideas in daily work will result in the learning activity being more effective. Training should be in line with what the organization needs in the short and long term.

There are also other training activities besides traditional courses. Other kinds of activities could be on-the-job training and practicing new and different ways of working. In this way, training will be built into ordinary work and be related to organizational goals.

Training can be more efficient if the REPI model described in Chapter 4 is used.

In summary, personal and social factors interact with each other. For example, reflection occurs in a group learning session, and with the right attitude, those in the group can make a heterogeneous environment more effective from a competence generation perspective.

Also, researchers such as Cepeda and Vera (2007) and Szulanski (2000) highlight that personal factors such as problem solving and attitude have an impact on the generation of new competences and emphasize that trial and error, learning by doing, and experimentation are important parts of the learning process. In addition, having a positive attitude toward work is required to achieve high performance (Spencer at al., 1994). The factors that have an impact on generating new competence are summarized in Table 1.1.

Table 1.1 Organization and Personal Factors That Support Generating New Competencies

Organizational level	Sharing	Sharing ideas and proposals. Discussing solutions and working together. Obtaining information from different parts of the organization. Helping others by sharing your knowledge.
	Social context	Stimulation to work, be seen and obtain feed-back. An innovative environment that inspires people to take responsibility. Feeling trust in the team and in the organization. Being allowed to make mistakes and have the freedom to learn. Supportive environment.
	Group learning	People who work and learn in groups. Discussing solutions and how to reach the goal. Working with a good team composition to create an environment of group learning. Learning from others in the team. Sharing ideas spontaneously.
	Heterogeneous environment	Bringing different competences together. Cross-functional projects. Interest in other areas. Working with business and clients.
	Time to learn	Time and space to seek information and learn.
Personal level	Attitude	Willingness and ambition to develop oneself. Attitude to do the best. Engagement and having an interest in the subject area. Motivation and taking responsibility. Being curious and having an interest in learning new things.
	Problem solving	Being exposed to new and challenging work tasks and situations. Solve a problem. Trial and error. Problems that one does not know how to address in the beginning. Testing different ways of doing things.
	Reflection	Following up and reflecting on how tasks were solved and learning from mistakes. Reflecting on experiences and on what to avoid or redo.
	Taking responsibility	Getting a new job or new task where you have to take responsibility.
	Training	Formal training and courses. Training in line with what the organization needs.

Another finding is that in organizations that are highly dependent on specialists and experts, personal factors become more important, whereas in organizations with a high degree of team culture, the social factors become more significant. A specialist has a tendency to only participate in his or her area of expertise and not be a part of the team as such. In his definition of team roles, Belbin (2012) defined a role called "Specialist" that, according to Belbin, involves only contributing in the short term and ignoring factors outside their own area of expertise.

It can be concluded that there is a negative correlation between specialization and learning. High levels of specialization impact negatively on learning, and specialists tend to be less engaged in group learning and knowledge sharing, which are important factors for learning. This finding is in line with Cabello-Medina et al. (2011), who argue that knowledge, skills, and expertise tend to be depleted over time. The willingness to share will also decrease in a specialist culture (Starbuck, 1992). But organizations need specialists and experts, and need to develop deep knowledge in areas that are core for the organization. Knowing that specialization has a negative impact on learning, managers can work to involve specialists in activities in which they can share their knowledge while also understanding other people's competence and creating a common view of goals and directions.

This section linked together competence with factors that have an impact on generating new competence, which in fact is learning. The next section will bring competence into different contexts and describe how the context impacts on the application of competence.

1.4 Context

Application and renewal of competence are influenced by the context in which competence is applied. What is meant by *context*?

Context can be seen from different perspectives. One perspective is the role that forms which kind of competence attributes need to be used. Those who work as salespeople need to be skilled in sales and know about the service or product that they sell. This is their knowledge. Apart from knowledge, they need different personal capabilities, such as being good at dealing with customers, patient, determined to finish a deal, having a positive attitude, and other personal capabilities. Of course, they need to have great social capabilities, such as being able to listen to the customer, build good relationships, etc.

If they instead work as system developers for new products, they need other kinds of knowledge, such as in technology, programming languages, etc.; they need other kinds of personal capabilities than the salespeople do, maybe being good at problem solving, innovative, and good at seeking information. Also, social capabilities may be different. System developers need to be good at working with the team and establishing networks with other system developers, but may not need to be as customer focused as the salespeople. So the role impacts how workers can apply their competence, and thus one needs to have different skill sets depending on their role.

Context could also be the structure of the organization. Is it a large, global, well-known company or a smaller, niched firm? In a larger company, the stakeholder situation can be more complex, with different national cultures and languages. In such a context, other kinds of competence attributes are necessary than in the smaller and local context. In the smaller company context, knowledge about the local market is important, whereas managing inter-cultural communication is more important in the multinational context.

Context can also be different at the company level. One example is the salesman who had worked several years in a large, well-known, and market-leading company in the telecom area. He was number one in sales and famous for signing the largest contracts in the company. The salesman was recruited by a new start-up firm within telecom. The start-up firm had a brilliant idea and very skilled developers, but they had no customers. One year later, the start-up firm still had no major contracts. The salesman failed to get any new customers or contracts. Why? The answer was simple. Coming from the well-known and market-leading large company, all doors were open and sales activities were about discussing price, delivery time, service-level agreements, etc. In the start-up firm, the sales activities were about contacting potential customers and trying to book a first meeting. The context for the salesman was totally different, and his competence was not in making cold calls and getting new customers.

Other aspects of context are industry, market maturity, level of knowledge intensity, heterogeneity, project maturity, specialist culture, etc. All context parameters impact the application of competence in different ways. A person who is successful in one organization will not automatically be successful in another organization.

If we look at context from a project management perspective, we can see that a project manager for a product development team needs different competences than a project manager managing several suppliers. In the product development example, the project manager needs to be good at team building and securing product requirements, whereas in the multiple-supplier example, the project manager needs to be good at vendor management and stakeholder management. Attributes such as industry, type of project, size, time, task, complexity, uncertainty, and stakeholder situation will have an impact on which kind of competence a project manager needs to have in a particular project.

Another aspect from a project management perspective is the strength of different competence dimensions in the specific context. We have the example of a project manager in a large company in the telecom sector. It was actually the same company as for the salesman mentioned above. The project manager started his career as a system developer in this specific area and continued as system architect and finally as project manager. He was very successful in his role as project manager, and the team delivered according to plan. In addition, he was highly respected, both by the project team and within the organization.

An adjacent technical area started to have problems with deliveries, and project management was not working in a proper way. The management team for the department decided to move the systems department project manager to the adjacent area to get structure and improve project management. The project manager failed in his new position. Why? The answer was that he had grown within his original area and had a deep knowledge of the system solution as such and could manage the project based on his deep technical knowledge. In his new role, he had no technical knowledge, and his weaknesses in other dimensions of competence became obvious. In this example, the context gave the project manager the possibility to be successful in one context but not in the other.

There are different areas of context dependencies for the project management area that can be important. As an example of context dependency, Reich, Gemino, and Sauer (2008) bring the concept of knowledge into the project context and propose four kinds of knowledge that are important, especially in IT projects:

- The first is *process knowledge,* which is the knowledge that project sponsors and team members have about project processes.
- Next is *domain knowledge,* which could be business, technical, or product knowledge.
- The third kind of knowledge is *institutional knowledge,* which pertains to the organization's values, power structure, and history.
- Finally, the fourth type is *cultural knowledge,* which can be described as the knowledge the project manager needs to have to manage teams composed of project team members from different areas and cultures.

Reich et al.'s work confirms that a project manager needs different competences depending on which project he or she will manage.

We can conclude that the context in which a person acts will impact the usage and renewal of competence. One important dimension of context is not mentioned in this section, namely *organizational culture and identity.* Because organizational culture and identity have a significant impact on competence and learning, it will be treated separately in the next section.

1.5 Organizational Culture and Identity

Organizational culture and identity could be considered as a context but, in this book, will be treated separately. The reason for treating this separately is that, when studying different organizations, it was evident that organizational culture and identity have high impact on both the application of competence as well as how the organization manages and develops new competence.

Culture in an organization is how a group shares values, beliefs, goals, and expectations (Reilly, 2012) that will persist over time, even when group members are changed (Kotter and Heskett, 1992). The culture will influence how people in the organization behave (Flamholtz and Randle, 2011). Identity, on the other hand, according to Melewar (2003), is related to how the organization allows itself to be known through a set of meanings, which also allows people to describe and remember it. The identity of an organization can also be described as what it expects to be and what it stands for (Garri, Konstantopoulos, and Bekiaris, 2013).

One of our case studies symbolized an organization with a dysfunctional corporate culture that made people avoid making decisions, tended to focus on operational issues instead of looking forward, avoided taking responsibility, and tended to blame others. The consequences of the dysfunctional corporate culture were unstructured ways of working, difficulties implementing decisions, delayed or unfinished projects, and inefficiency. In contrast, another of the case studies showed a strong, functional corporate culture in which management and employees acted in line with corporate values. This particular organization was the Research and Development (R&D) department in a large, fast-growing high-tech company.

A strong, functional corporate culture was described by Flamholtz and Randle (2011) as a culture that supports corporate goals, which people can clearly articulate and understand. In the case of a strong and functional organization, the R&D department still has some barriers to supply chain operations, marketing, and the IT department, which its managers think have another interpretation of the corporate values.

This sheds light on the fact that different professions may have different cultures to which people feel an identity. Professional identity is created when one learns about the work related to the profession, and also when one acts in a social environment with others in the same profession who give implicit feedback about one's performance, which is a form of social validation of professional identity (Pratt, Rockmann, and Kaufmann, 2006). According to Pratt et al., change in identity is strongest when people's work does not match who they are as professionals. Different professions develop their own language and way of working. As a project team member in one of the case studies expressed it: "They develop their own language with their own terms and concepts, and are using different vocabulary than what we are." A project team member in the same case study explained it as: "We also have IT versus business, and we do not understand each other."

We can conclude that people feel identity in their professions. This conclusion is in line with Hofstede (1998), who argues that an organization can have different subcultures that depend on their working tasks. He also points out that higher management needs to be aware of the variety of different professional cultures within the organization, especially in complex organizations.

One of the companies in the multiple-case study acted in a global environment, with the organization spread over the globe. One of the project managers expressed the national cultural differences in the following way: "We suffer from being global and localized in different parts of the world. There are so many cultural differences, everything from project management philosophies to how to conduct a meeting." It is obvious that national culture impacts on how competence can be applied.

Another cultural dimension that affects competence management is the industry culture. The purpose of the three different case studies was to look for similarities and dissimilarities between a public-sector organization and privately held organizations, but also between a growing company and a declining one. During the study, it became clear that different industries also develop different cultures.

The public-sector organization is also a regional transport authority. The values and behavior in the public-sector organization have more in common with other regional transport authorities than with the health-care organization that is the major part of the public organization. People working with public transport identified with people working with other actors in the public transport industry. These different relationships were observed during the case study.

In the studied organizations, it could also be observed that the IT departments at the different companies shared several common behaviors, showing that the IT departments acted in the same way in relation to different business units. The case studies showed that people felt an identity with their profession, as described by Pratt et al. (2006), who argue that professional identity is formed around an organized group that possesses unique knowledge and skill sets, such as those in law, medicine, and IT.

In summary, organizational culture and identity can be seen from four different perspectives—corporate, national, profession, and industry—as depicted in Figure 1.4. All of these perspectives have an impact on the application and development on competence.

Having the four different components in mind, we can see that an organization can develop different subcultures. One example, as mentioned above, is that IT departments in many organizations develop their own subculture with their own language and behavior. The IT culture also makes it easier for people moving between IT departments in different companies. We can see the same for finance, supply chain, and other functional areas. An interviewed project team member expressed his view of different cultures within the organization as: "They are using their own strange phrases, and we do not understand what

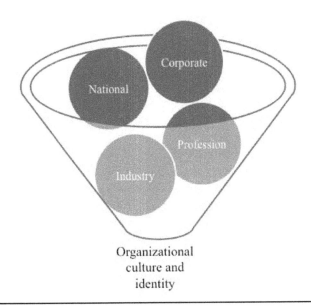

Figure 1.4   Organizational culture and identity.

they mean." The project manager for one business-oriented IT project put it as: "We are struggling to define the requirements because they are not thinking like us."

Being aware of organizational culture and identity, we can also create subcultures to establish a platform for working efficiently and for learning. For example, a project manager can establish a subculture in the project by highlighting the norms, behaviors, and paths to communicate in the specific project. In this way, the project manager can establish a positive culture, even if the organizational culture to some extent is not positive and functional.

One interviewed project manager described how he established the common culture in this way: "I wrote a 'working-in-the-project' document with values and statements like: avoid sending emails, call for short meetings when needed, talk to each other, etc., then I worked with these values in the project to establish a culture of solving problems. Project communication was the major issue in the previous project in this area."

However, it is very difficult to establish a positive and functional project culture if the company culture is dysfunctional. We tend to bring with us the negative effects from the dysfunctional culture into the projects.

In this section, we viewed competence from different dimensions, and we also concluded that the context has a significant impact on competence, which will be further analyzed in the next section.

1.6 The Relationship Between the Six Dimensions of Competence and Context

If competence is context dependent, how are the different dimensions of competence related to the context?

Competence as such is individual, as emphasized by Le Deist and Winterton (2005), and if competence is individual, is it the application of competence that makes it dependent on the context?

- As previously explained, *personal capability* is the personal characteristics a person has, and also one's attitude to work and ability to use knowledge to solve a problem or perform a task. This dimension is related to the personal capability to use knowledge. This is actually what Hager and Gonczi (1996) mean when they argue that competence is context dependent. Here we can see that in an organization with mistrust or a culture of not allowing people to make mistakes, a person with low self-confidence will have less possibility to apply his or her knowledge, compared to an organization based on trust and fostering possibilities to try new and unknown things.
- Another dimension of the competence concept is *knowledge and experience,* which was described in the case studies as the knowledge an individual has based on education, previous work, or other sources. Knowledge is also related to a subject matter area, such as knowledge about a specific technology, business, sales, or other similar areas. This dimension is task dependent, because the individual needs to be knowledgeable in the subject matter area related to the working tasks, which in turn can be considered as related to a role. It can also be the case that the need of technical knowledge for a role in one organization is different from the need in another, depending on different kinds of prerequisites.
- The *social capability* dimension was shown to have a direct relationship to the organizational culture. The organizational culture impacts, among other things, how we communicate with each other. Guiso, Sapienza, and Zingales (2008) stated that people learn from others' experience, and that if they do not trust others they will not trust the information that others share, and thus they will not learn from others. In the case of a dysfunctional organizational culture, in which people mistrust each other, a person with low social capability will have more difficulty collaborating with others than in an organization that is based on trust and togetherness.
- The *ability to learn* dimension is related to how an individual learns new things by taking in information and interpreting new knowledge. Previous studies showed that organizational culture has a significant impact on learning, because the organizational culture determines values

and beliefs that encourage learning and knowledge sharing (Liao, Chang, Hu, and Yueh, 2012). Learning is more effective when we apply new knowledge, which means that in a context in which we are allowed to experiment, try new and unknown things, and elaborate on problems, the ability to learn will be higher.

- The last dimension, *ability to manage complexity*, is related both to an individual's capability to manage a complex environment, link different domains, and participate in complex problem solving, as well as to the complexity of the context. For example, in a complex stakeholder situation, we need to have higher ability to manage the complexity than in a context with few stakeholder interactions.

These examples show that the context has an impact when people apply their competence, and also that the different dimensions in the competence concept are affected in different ways.

In the next section, the six dimensions of competence, the context, and the organizational culture and identity will be merged together into a competence concept called "The Competence Lemon."

1.7 The Competence Lemon

As seen above, the derived competence concept consisted of six dimensions—namely, *personal capability, knowledge and experience, social capability, ability to manage complexity, ability to learn,* and *leadership qualities.* We also concluded

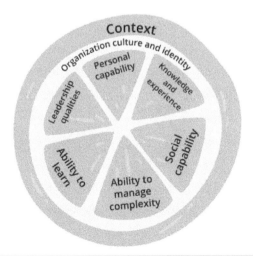

Figure 1.5 The competence lemon. *[Figure is described on the following page.]*

that the context and the organizational culture and identity have an impact on the application of competence. Based on this reasoning, the competence concept in Figure 1.5, called the *competence lemon,* can be outlined. Figures 1.2 and 1.3 gave a summarized description of every dimension of competence.

Although the context and the organizational culture have an impact on the application of competence, the weight of the six dimensions varies depending on in which context and organizational culture the competence is applied. For example, a project manager could be successful in one context but less successful in another, depending on how strong he or she is in certain competence dimensions. If the project manager leads an internal product development team, he or she needs to be strong in motivating the team to perform, whereas in managing a change management project with many stakeholders, the project manager needs to have high social capabilities.

Having defined the concept of competence, the next chapter will go into how an organization can benefit from effective competence management, and what mechanisms constitute competence management.

Chapter 2

The Competence Loop – A Framework for Efficient Competence Management

In the previous chapter, we examined the concept of competences, consisting of six different dimensions, which in turn are influenced by the context and the organizational culture in which the competences are applied. In this chapter, competences are brought into an organizational context to determine how they can be managed in such a way that they contribute to an organization's innovative capabilities and competitive advantage. All references to the competence lemon can be found in Chapter 1.

We start by identifying the concept of core competences and competence management and continue by investigating the dynamic capabilities that link competence management to a competitive advantage.

2.1 Core Competences

Subramaniam and Youndt (2005) emphasize the close connection between an organization's ability to innovate, its intellectual capital, and its ability to utilize its knowledge. Intellectual capital can be seen as human capital in conjunction with social capital and organizational capital (Lengnick-Hall et al., 2009; Subramaniam and Youndt, 2005). Social capital in this context is represented by individuals' ability to share and exchange knowledge insights and mental models (Cabello-Medina et al., 2011). Cabello-Medina et al. (2011) also argue that employees with highly valued knowledge and skills contribute to innovation, because they are more willing to experiment and apply new ways of working. Teece, Pisano, and Shuen (1997) emphasize that an organization's greatest potential contributions to strategy are skill acquisition, learning, and the accumulation of intangible and organizational assets.

In addition, human resource management (HRM) practices could support the creation of a competitive advantage by aligning the practices with the firm's strategy. In doing so, human capital could achieve a greater sustainable competitive advantage if an organization adopts those procedures and practices that promote the enlargement of the knowledge value for the organization (Lopez-Cabrales, Real, and Valle, 2011).

Moreover, the human resource advantage can be seen from two perspectives—human capital and organizational processes; the human capital advantage stems from having more capable people than the competition, and the organizational process advantage is achieved by having more effective working procedures than one's competitors (Lengnick-Hall et al., 2009). From a competence management perspective, one interpretation of the above-cited work could be that an organization should have more effective competences and competence management procedures than its competitors to achieve the human resource advantage.

As seen in the previous chapter, knowledge is the part of competences to be applied, and the other dimensions of competences are facilitators either of the accumulation of new knowledge or of the application of existing knowledge in a high-performing way. From an organizational perspective, knowledge can be stored in three main repositories—individuals, tools, and tasks—of which tools are technological components and tasks reflect the goals, purposes, and intentions of the organization (Argote and Ingram, 2000).

As described in the previous chapter, human competences are individual but can also be linked to the organizational level. Core competences, on the other hand, can be seen as organizational capabilities that create a competitive advantage for firms. Prahalad and Hamel (1990) define core competences as the collective learning within an organization as well as the coordination of production skills and the integration of different streams of technology. Furthermore, core competences are those capabilities that are critical to a business in achieving a competitive advantage, are built on continuous improvements and enhancements (Eden and Ackermann, 2010; Hamel and Prahalad, 1994), and are manifested in business processes and activities (Agha, Alrubaiee, and Jamhour, 2011).

The relationship between core competences, competitive advantages, and organizational performance is successfully tested empirically by Agha et al.

(2011). They show that core competences are a significant determinant of organizational performance and competitive advantages in the sense that more competences lead to a higher degree of organizational performance and competitive advantage. Furthermore, they argue that responsiveness and flexibility are two dimensions of a competitive advantage and that these are antecedents to organizational performance. Their findings show that the ability to manage the complexity dimension from the competence lemon (page 27) needs to be well developed by people in the organization to achieve a competitive advantage.

Distinctive core competences should preferably be maintained and sustained; otherwise, they could dissolve or no longer be valid for a competitive advantage (Agha et al., 2011; Eden and Ackermann, 2010). To outperform the competition in the long run and achieve a sustained competitive advantage in the market, an organization's leaders need to define its core competences (Clardy, 2008; Laakso-Manninen and Viitala, 2007), which can be achieved by analyzing the relationship between the organization's competences and its purpose and business goals (Eden and Ackermann, 2010).

One way to find the link between an organization's business goals and its competences, which defines the organizational core competences, is by describing the organization's potential successes and failures (Eden and Ackermann, 2010). Gupta et al. (2009) include learning as a core competence in a knowledge-intensive context.

Chen and Chang (2010) describe a model in which competences are described with visual and hidden attributes. Individual competences have skills and knowledge as visible attributes and self-concept, motives, and traits as hidden attributes, whereas organizational (core) competences have strategic skills and strategic knowledge as visible attributes and organizational image and strategic intents as hidden ones. They also emphasize that visible attributes have high strategic value, whereas hidden attributes have low strategic value. In addition, visible attributes have low uniqueness, whereas hidden attributes have high uniqueness in their model.

The organizational image is the view that the employees have of the organization—its perceived organizational identity.

The above reasoning connects core competences to human resources as a source of the sustained competitive advantage that is crucial for a company's success. However, core competences are not solely human for an organization; they are based on and consist of knowledge—that is, human, social, and organizational capital embedded in people and systems, combined with the creation, transfer, and integration of knowledge (Wright et al., 2001). Individual knowledge (part of human capital) becomes institutionalized and codified (organizational capital), is transferred between people through networks (social capital), and forms an organization's intellectual capital (Subramaniam and Youndt, 2005).

The organizational context facilitates the relationship between human and core competences. Attributes such as mutual trust, shared values, vision, and strategy are the bases for sharing the same mindset in understanding and reaching the goals of an organization (Chen and Chang, 2011).

To be innovative and achieve a competitive advantage, the way of managing competences is crucial for an organization. According to Harzallah, Berio, and Vernadat (2006), competence management is a way for an organization to manage its competences at the corporation, group, and individual levels. They develop a three-step competence management model that consists of three basic processes: competence *identification,* competence *assessment,* and competence *usage.*

- **Competence identification.** The competence identification process aims to identify the competences required to perform strategic actions, missions, tasks, and similar activities.
- **Competence assessment.** The next process, competence assessment, focuses on deciding whether an individual has acquired a specific competence.
- **Competence utilization.** The last process, competence utilization, aims to decide how to use the information from the two other processes to achieve efficient utilization of competences.

Furthermore, they distinguish between *individual, collective,* and *company* core competences, in which collective competences are related to a group of people and core competences to the aggregated competence that sustains the company's competitive advantage in terms of products or services. The model brings in competences to three levels: individual, group, and organization.

Competence management is based on human capital, and Shih and Chiang (2003) make a connection between core competences, HRM practices, and corporate strategy, highlighting the need to design HRM practices to fit the desired employee competences. The HRM practices that they point out as core competences are *recruitment and promotion, training and development, performance appraisal,* and *compensation systems.*

Medina and Medina (2014) further develop Harzallah et al.'s (2006) and Shih and Chiang's (2003) works and define HRM competence management practices as being constituted by "selection," "training and development," "performance measurement," and "internal promotion." In their model, they link competence management to HRM and position HRM competence management practices as a subset of HRM practices. They also adapt the model to project-intensive organizations and show that HRM competence management practices are the mechanisms that link an organization's long-term goals in terms of competences with its work in projects, as depicted in Figure 2.1.

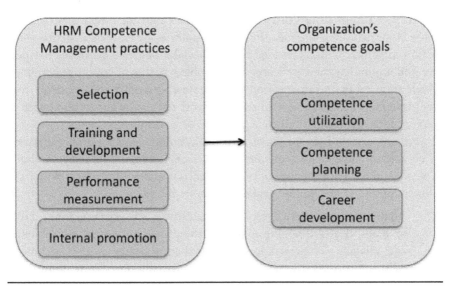

Figure 2.1 Relationship between HRM competence management practices and organizational competence goals.

After outlining the concept of core competences and linking it to competence management, we continue to examine dynamic capabilities, which are the link between competence management and competitive advantages.

2.2 Dynamic Capabilities

Dynamic capabilities have their roots in the resource-based view and in a firm's ability to integrate, build, and reconfigure its external and internal competences to meet changing market conditions and, in this way, create a competitive advantage (Eisenhardt and Martin, 2000; Peteraf, 1993; Teece et al., 1997; Wright et al., 2001).

In addition, dynamic capabilities refer to the ability to achieve new forms of competitive advantage—that is, how organizations can demonstrate timely responsiveness and respond to the market's need for product innovation in a rapid and flexible manner, combined with the management capability to coordinate and redeploy external and internal competences efficiently (Eisenhardt and Martin, 2000; Hubbard, Zubac, and Johnson, 2008; Killen, Hunt, and Kleinschmidt, 2008; Teece et al., 1997).

The difference between dynamic capabilities and processes is explained by Wang and Ahmed (2007), who refer to capabilities as a firm's capacity to deploy resources, encapsulating both explicit processes and tacit elements, such as know-how and leadership, meaning that capabilities are embedded in processes. Zollo and Winter (2002) consider dynamic capabilities from a learning perspective and argue that they are learned and stable patterns of collective activities, which are used in an organization to generate and modify the operating routines systematically to achieve improved effectiveness. The latter can be seen as a continuous approach for efficiency in ways of working.

Dynamic capabilities consist of three main components for an organization to be able to compete in a dynamic market environment—namely, *adaptive, absorptive,* and *innovative* capabilities (Biedenbach and Müller, 2012; Wang and Ahmed, 2007), which will be explained further below.

- **Adaptive capability.** Adaptive capability is an organization's ability to adapt in a fast-moving market through means of strategic flexibility and balancing its exploration and exploitation strategies. In this context, dynamic capability is shown by the ability to adapt to environmental changes and align the organization's resources with the external market demands, which is crucial to the organization's survival in the market (Biedenbach and Müller, 2012; Wang and Ahmed, 2007). This capability shows how the organization can understand and interpret the market in which it acts and adapt quickly to the changes.

- **Absorptive capability.** The next capability, absorptive, refers to the ability to absorb external information and knowledge, integrate it with internal knowledge, and apply it in a way that contributes to the organizational goals. High absorptive capacity means the ability to learn from others, integrate external information, and transform it into knowledge embedded in one's own organizational processes (Biedenbach and Müller, 2012; Jiménez-Barrionuevo et al., 2011; Volberda, Foss, and Lyles, 2010; Wang and Ahmed, 2007). External information can be absorbed from customers, partners, suppliers, and other actors and stakeholders.

  To develop a high absorptive capacity, an organization needs to have processes and procedures as well as a culture to capture external information and use it to develop new products, services, or processes to achieve a competitive advantage. From a project perspective, absorptive capacity can indicate how well the project team learns from a supplier and uses the knowledge in a current project or in upcoming projects.

- **Innovative capability.** Finally, innovative capability refers to the ability to align strategic innovative orientation with innovative processes and behavior, thereby developing new products, services, or markets (Biedenbach and Müller, 2012; Wang and Ahmed, 2007).

  An organization's ability to innovate is directly connected to its intellectual capital (Cabello-Medina et al., 2011; Subramaniam and Youndt, 2005) and is supported by efficient utilization of organizational knowledge through the generation of new ideas and the exploitation of existing human capital as well as the organization's ability to grow and progress in a changing environment based on the generation of new behaviors and ideas (Kocoglu et al., 2012). Furthermore, the capacity to innovate could be considered as a learning process (Oltra and Vivas-López, 2013), in which the most relevant feature is the uniqueness of knowledge (Cabello-Medina et al., 2011).

Innovative capabilities can be divided into two different capabilities, namely *incremental* and *radical* innovation (Lin et al., 2013; Subramaniam and Youndt, 2005; Tamayo-Torres et al., 2010), of which incremental innovative capability requires the reinforcement and exploitation of existing knowledge, whereas radical innovative capability requires the transformation of existing knowledge and the disruption of an existing trajectory (Biedenbach and Müller, 2012; Subramaniam and Youndt, 2005; Tamayo-Torres et al., 2010). In addition, innovation ambidexterity refers to instances in which an organization simultaneously achieves radical and incremental innovation (Eriksson, 2013; Lin et al., 2013).

An organization needs to work with both radical and incremental innovation to achieve a sustainable competitive advantage. Too much focus on incremental innovation could lead to the organization's becoming outdated. On the other hand, too great a focus on radical innovation could eventually lead to a situation in which the organization becomes bankrupt before it has received a large enough return on investment (Lin et al., 2013).

Eisenhardt and Martin (2000) also point out the knowledge creation process as a crucial dynamic capability, especially in knowledge-intensive firms. This is further elaborated by Eriksson (2014), who, in a review of 142 academic articles about dynamic capabilities, finds four vital knowledge processes supporting dynamic capabilities: *knowledge accumulation, knowledge integration, knowledge utilization,* and *knowledge reconfiguration and transformation.*

- **Knowledge accumulation.** The first process, knowledge accumulation, refers to knowledge being acquired through experience with two different objectives—namely, the replication or renewal of the existing knowledge—and balancing the two is a dynamic capability prerequisite.
- **Knowledge integration.** The second process, knowledge integration, connects new knowledge with existing knowledge by combining various resources. In this process, the accumulated knowledge becomes relevant to the organization.

- **Knowledge utilization.** The third process, knowledge utilization, is an often-neglected key process in which the organization benefits from accumulated and integrated knowledge by different types of knowledge sharing. These can be individual tacit knowledge sharing or, at the organizational level, explicit knowledge sharing.
- **Knowledge reconfiguration and transformation.** Finally, the fourth process is knowledge reconfiguration and transformation, in which the organization either generates new combinations of its existing knowledge or leverages its existing knowledge in new ways or for new purposes. Resource management, the organization's ability to transform knowledge resources, directly affects its ability to sense opportunities.

The three dynamic capabilities mentioned above are crucial for attaining a competitive advantage. To develop and maintain those capabilities, an organization needs something more. Learning capabilities can be seen as second-order capabilities, in that they have a role in the creation of other capabilities that are considered as first-order capabilities; this is due to the fact that second-order capabilities have the ability to change other capabilities (Killen et al., 2008). Lin et al. (2013, 262) define learning capabilities as "the combination of practices that promote intraorganizational learning among employees, partnerships with other organizations that enable the spread of learning, and an open culture within the organization that promotes and maintains the sharing of knowledge." Learning capabilities are based on tangible and intangible resources (Kocoglu et al., 2012) and support organizations in the exploration and exploitation of knowledge through different flows from the individual level to the group and organizational levels (Li, 2012).

Learning as such can be considered as a permanent change in behavior based on experience and reception, which leads to better performance (Gunsel et al., 2011; Holmqvist, 2003), as well as an organization's ability to acquire resources by the acquisition of new resources or knowledge on how to use their existing resources (De Wever, 2008). Szulanski (2000, 12) elaborates on the learning process by referring to the experimentation or planning stage before the actual use of knowledge as "learning before doing," and the phase during which knowledge is put to use, entailing the resolution of unexpected problems, as "learning by doing." Here we can connect to projects in which the planning phase is related to "learning before doing," while "learning by doing" is carried out in the execution phase. Those learning processes are based on both factors on the personal and organizational level, as described in Chapter 1.

Another view of learning is the "learning by working" approach, because learning can be seen as knowledge creation through social participation in everyday work—for example, in project teams transcending organizational boundaries (Fenwick, 2008). This is in line with Holmqvist (2003), who states that learning is a social activity in which individual learning takes place in social contexts. The "learning by working" approach mainly contains factors on an organizational level that exert an impact on the generation of new competences, as described in Chapter 1.

Viewing learning from an organizational perspective, we can see that it is a rather spontaneous or informal process (Oltra and Vivas-López, 2013) of acquiring, transferring, and integrating new knowledge and, in this way, adding value to the organization (Jerez-Gómez, et al., 2005).

Organizational learning enables an organization to develop capabilities that support innovation, which in turn has a positive effect on performance. Innovation occurs when individuals share knowledge that generates new and common insights, leading to organizational innovation (Jiménez-Jiménez and Sanz-Valle, 2011). In this process, social capital has an indirect influence on innovation through human capital (Cabello-Medina et al., 2011) and on organizational performance through knowledge transfer (Maurer, Bartsch, and Ebers, 2011). Oltra and Vivas-López (2013) elaborate further on this theme, arguing that innovative capacity is a learning process based on knowledge, and one in which experimentation and openness are organizational capabilities that require constant interaction and social networking.

Moreover, organizational learning can be divided into two different types: *exploratory* and *exploitative* learning (Holmqvist, 2003; Kang, Morris, and Snell, 2007; Kang, Rhee, and Kang, 2010). Exploratory learning is symbolized by the integration of external knowledge, which does not currently exist in the organization, with the creation of new value for the customer, or the replacement of existing knowledge to improve the current customer value; exploitative learning involves the expansion of existing knowledge to enrich the current customer value (Brady and Davies, 2004; Kang et al., 2007, 2010).

Nevertheless, there is a risk that an organization with too great a focus on exploitative learning could reach a state in which the knowledge base becomes obsolete—although the organization utilizes its knowledge stock, it fails to renew it (Kang et al., 2007, 2010; Lin et al., 2013). There is also a risk that those organizations that focus on short-term returns (Brady and Davies, 2004) based on the continuous exploitation of existing competences, without a process of renewal, end up falling into the competence trap (Eriksson, 2013). On the other hand, organizations that are too focused on the exploration of new possibilities could suffer from too few competences and too many unexplored ideas (Brady and Davies, 2004) and eventually fall into a vicious trap in which failure leads to more research followed by change, which leads to more failure (Eriksson, 2013).

Based on the above reasoning, it can be concluded that learning is a second-order capability, as described by Killen et al. (2008), and it has a relationship

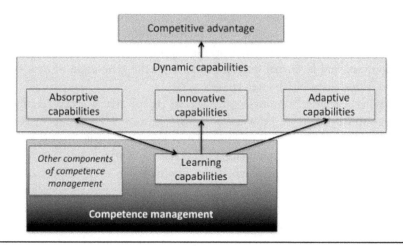

Figure 2.2 The link between dynamic capabilities, learning, and competence management.

with competence management and the ability to change the innovative, adaptive, and absorptive capabilities, as shown in Figure 2.2.

Having linked competence management to competitive advantages through dynamic capabilities, we will move the competence management concept into the project context.

2.3 Competence Management in Project-Intensive Organizations

A *project-intensive organization* is one in which functional organization and projects coexist, and a considerable number of activities are conducted in projects. This is the situation in many organizations today. Projects are used to develop new products or services, drive change, establish the firm in new markets, purchase new solutions, and many other purposes. In this kind of organization, interactions take place between the permanent organization and the different projects executing different kinds of activities. Projects can, in many cases, be seen as a competence arena in which new competences evolve. I will further elaborate on the concept of project-intensive organizations in Chapter 3.

Koskinen (2009) argues that project-intensive organizations must be able to acquire new knowledge and skills continuously to be successful in their markets. Furthermore, a high degree of social capital improves the capabilities and skills of individuals. In project-intensive organizations, knowledge sharing and learning have central roles and are the most important capabilities and bases for innovation (Artto and Kujala, 2008). Furthermore, it can be stated that different kinds of project-intensive organizations have different kinds of weaknesses in sharing, transferring, and integrating knowledge between projects and between the projects and the parent organization (Brady and Davies, 2004; Hobday, 2000; Pemsel and Müller, 2012). Activities such as lessons learned, retrospective meetings, hand-over procedures, and so on are implemented in different ways in different organizations.

2.3.1 Project Capability Building

Project capability building was introduced by Brady and Davies (2004), who define two co-evolving and interacting levels of learning—namely, *project-led* and *business-led* learning. They describe project-led learning as bottom-up exploratory learning that occurs when an organization moves into new markets or utilizes new technologies. Project-based learning is divided into three phases.

Project-Led Learning

- **Within-project learning.** The first phase is the establishment of a new exploratory project, and the focus is within-project learning. Popaitoon and Siengthai (2014) explain within-project learning as occurring within situations in which projects leverage the absorbed knowledge to develop specific tasks, and they connect this absorbed knowledge to a project's realized absorptive capacity. These kinds of projects could be considered as exploratory projects in which new, radical innovations are made. The learning focus is within the project to acquire new knowledge.
- **Project-to-project learning.** In the second phase, competences are transferred from the vanguard project to subsequent projects in which they can be used, and the focus is on project-to-project learning. This phase is what Popaitoon and Siengthai (2014) call the project's *potential absorptive capacity*—when the cumulative knowledge from prior projects is transferred to subsequent projects. The absorptive capacity in this context is seen from the project's point of view—that is, the project absorbs knowledge from the previous project.
- **Organizational learning.** In the third phase, competences are transferred to the parent organization. Transferring knowledge from projects to the parent organization can be considered as organizational learning, whereby the knowledge is captured and stored within the organization. Based on this reasoning, it is proposed that a project's capacity to absorb, use, and transfer knowledge could be called *project-intensive learning.*

Business-Led Learning

Learning in the other direction, from the parent organization to projects, is referred to as business-led learning by Brady and Davies (2004). The aim of business-led learning is to use the competences from projects and refine and extend the organization's routines and capabilities to exploit fully its new market and technology base. This type of learning is based on top-down strategic decisions that aim to create and exploit the resources and capabilities in an organization to perform foreseeable and routine project activities. Based on Brady and Davies' work, Bredin (2008, 566) adds *people capability* as the knowledge, experience, and skills embedded in procedures and routines that are important in project-intensive environments. People capabilities focus on the capacity to access, develop, and maintain a workforce over time, independent of turnover.

Based on the project capability concept of Brady and Davies (2004) and the people capability concept of Bredin (2008), Melkonian and Picq (2011) connect the strategic perspective with learning evolving through projects in a double-loop process. In the top-down process, organizational routines support project performance through HRM practices and policies for the selection and preparation of teams and team members to work on projects.

The bottom-up process aims to secure the contributions of different projects with respect to the constant evolution and changing of the organization through feedback experience, project-based learning, the constant improvement of HRM policies, and the readjustment of individuals' career and development plans. Comparing the work of Melkonian and Picq (2011) with Medina and Medina's (2014) HRM competence management practices, we can see similarities in the ways in which resources are selected for projects, competences are measured, training is performed, and project achievements are linked to career development.

2.4 The Competence Loop

By connecting dynamic capabilities with learning capabilities and putting them into a project-intensive context that also considers the interaction between a project and its parent organization, the framework in Figure 2.3, called *the competence loop,* can be drawn.

The first part of the framework is based on the four mechanisms in the competence loop:

- Competence *utilization* is the mechanism for utilizing the existing competences in accordance with the organization's strategic goals through resource management and formal training, and it is the interface between the parent organization and the project.

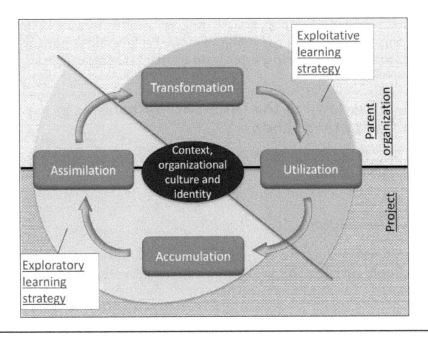

Figure 2.3 The competence loop.

- Competence *accumulation* is the mechanism for creating or acquiring new competences through the work in projects or the absorptive learning that takes place in the project. From the project perspective, competence accumulation can also be related to a project's realized absorptive capacity, as described by Popaitoon and Siengthai (2014).
- Competence *assimilation* is the mechanism for assessing, interpreting, and understanding new knowledge in relation to core competences and is the interface between the project and the parent organization.
- Finally, competence *transformation* is the mechanism for performing competence planning and combining new and existing competences to identify competence gaps, update the knowledge base, or reconfigure core competences.

The second part of the framework is based on the organization's learning strategy. An *exploratory learning strategy* can be seen as the organization's ability to generate new competences and interpret them, while an *exploitative learning strategy* is the organization's ability to combine newly generated competences with existing competences and utilize the combined competences in new projects. The learning strategies should be balanced between exploratory and exploitative learning to make a long-term contribution to the competitive advantage.

In the same way as for competences, the usage of the competence loop is also influenced by the context and the organizational culture and identity. We will examine this impact in greater depth when analyzing the different mechanisms.

2.5 Breakdown of the Different Mechanisms in the Competence Loop

This section will explain the different mechanisms in more detail and break them down into factors.

2.5.1 Utilization

The first mechanism is the *utilization* mechanism, which concerns the way in which competences are utilized in line with the strategic goals of the organization.

Competence Allocation

Regarding the different factors of the utilization mechanism, the first factor is *competence allocation,* which refers to the way in which competences are allocated to different projects and depends on the previous identification of the competences needed to fulfill the business goals and project needs. This factor focuses on the allocation of competences rather than on the traditional resource allocation. To meet the project goals, the project needs the right competences, rather than project roles that are filled with resources without considering their competences based on the competence lemon.

Different organizations manage the resource and competence allocation in different ways. In many organizations, the project manager identifies the competences that the project needs, and the functional manager is responsible for allocating the resources, but the method can differ between companies. The major change in moving from allocating *resources* to allocating *competences* is the view of people based on their competences and not their role. A senior manager in one company explained their situation as follows: "We focus more on using our resources than even thinking about what competences we need in the project," which in their case led to a low utilization of competence when the focus was on resources and not competence.

A project needs competences to work on project tasks to fulfill the project goals. One example is a project undertaken to deliver maintenance releases for a technical product in the telecom market. The project had a rather low priority in the organization, and the resources allocated to the project had neither knowledge and experience of the technical platform nor the competences needed to tackle the technological challenges. Consequently, the project did not deliver any solutions, and the dissatisfaction increased.

The management team recognized the low performance and allocated one new resource to the project—a person who was very skilled in the technical platform and had long experience gained from working closely with the customer and training on the solution. The new resource had knowledge and experience not only of the technical platform but also of working with customers, pedagogical skills (personal capabilities), and a high level of social capability. The role of the new resource was to support the project team, train the team members, and enable them to develop and deliver. After a few weeks, the project slowly started to deliver new releases, and the resources' competences started to increase. In this case, resources without the right competences were allocated to the project, leading to low project performance.

In one studied company, one of the project managers expressed the competence issue in the following way: "It happens that we do not have the resources or the competences we need. The consequence is usually a delayed project, but normally we resolve it by bringing in a consultant, move a resource from another team, or reschedule the project."

Project Portfolio Planning

The next factor is *project portfolio planning,* which describes the way in which the portfolio of strategic projects is organized. Project portfolio planning has an impact on the project outcome through the method of project initiation and control. Moreover, it has an impact on competence utilization, because project portfolio planning provides an overall picture of how competences are allocated to the different projects. Without effective project portfolio planning, organizations have tendencies to start projects without having control of the scope, requirements, or competence needs. With a competence perspective on a project portfolio, competences can be distributed where they are needed through the competence planning factor described below. A project portfolio has a priority dimension that facilitates competence planning and allocation.

New Resource Introduction

The factor *new resource introduction* refers to the way in which new resources are introduced into the work efficiently. Normally, the functional manager is responsible for introducing new resources to the organization, but, in a project-intensive context, the team will in many cases take over the practical introduction. A functional manager for a department working with agile project management (Scrum) expressed his way of introducing new resources: "Into the Scrum team, they will take care of him or her." The combination of the use of agile project management methodologies and a culture in which processes and documentation have low priority, the introduction of new resources is made based on tacit to tacit knowledge transfer, or what Nonaka (1994, 19) calls *socialization*. An organization with a high resource turnover is more often dependent on onboarding documentation to introduce new resources.

Recruitment

The next factor, *recruitment,* describes the action of employing new co-workers and is correlated with the competence concept. In many cases, recruitment activities focus on the knowledge and experience dimension of competences, while, in a knowledge-intensive environment, all the dimensions of the competence loop should be considered. One of the interviewed HR business partners expressed her view of recruitment in the following way: "Usually, I am looking at how well the candidate fits into our culture, but also if the candidate has the right passion and burn for his or her work." This organization has a strong functional culture in which the organizational culture is always present. A person with a high ability to learn, social capabilities, and personal capabilities probably has greater potential to take on different roles in the organization and be a high performer.

Competence Planning

Another factor in the utilization mechanism is *competence planning,* which refers to the way in which responsible people in the organization identify and plan the kinds of competences that are needed to execute the projects in the project portfolio. In this factor, both team composition and individual competences need to be considered to balance the project portfolio with the strategic goals. The project portfolio is the input to competence planning, and resources with the right competences are allocated to projects that need their competences, based on the organizational goals and priority.

Sourcing of External Competences

The factor *sourcing of external competences* is the process of selecting the external workforce or evaluating suppliers to fill the competence gaps. The reasons for acquiring external competences can differ—for example, resource gaps or a lack of competences in the organization. Regarding recruitment, all the dimensions of the competence lemon should be considered when using an external workforce, especially the fit with the organizational culture and the team. In many cases, external resources are selected based on their knowledge and experience rather than on how well they fit into the project team. The risk is that they will have a negative impact on the team spirit and performance if not all the dimensions of the competence lemon are considered when selecting them.

Project Manager (PM) Competence

The last factor in the utilization mechanism is *project manager (PM) competence.* This factor relates to the kinds of competences that a project manager needs to manage different kinds of projects—for instance, IT infrastructure management, change management, or a more business-oriented project. Successfully managing a project in one context does not automatically mean that the project manager will be successful in another context. In addition, for the selection of the project manager, all the dimensions of the competence lemon should be considered. Selection of the right project manager is further discussed in Chapter 6.

The factors constituting the utilization mechanism are summarized in Table 2.1. The factors in Table 2.1 are not weighted, although the context and the culture

Table 2.1 Factors Constituting the Utilization Mechanism

Factor	Description
Competence allocation	The process of allocating competences to projects linked to the strategy or organizational goals. Focus on competences and not only resources.
Project portfolio planning	The process of organizing strategic activities into a portfolio of projects.
New resource introduction	How new resources are introduced into the work in an efficient way.
Recruitment	How the recruitment of employees is linked to the organization's strategic goals. Recruitment considers all the dimensions of the competence concept.
Competence planning	The process by which the organization identifies and plans the kinds of competences that are needed to execute the projects in the project portfolio. Input for the competence allocation factor.
Sourcing of external competences	The process of selecting the external workforce or evaluating suppliers to fill competence gaps.
Project manager competence	The kinds of competences that a project manager needs to manage different kinds of projects.

exert an impact on the importance of the different factors and the way in which they are implemented and managed in the organization. For example, the factor sourcing of external competences is more important in an organization that is heavily dependent on external resources, whereas the factor recruitment is more important in an organization with significant organic growth.

Having described the utilization mechanism, we proceed to the next mechanism in the competence loop—accumulation.

2.5.2 Accumulation

The next mechanism is the *accumulation* mechanism, which concerns how the organization creates or acquires new competences. The accumulation mechanism is closely related to the factors that have an impact on generating new competences, described in Chapter 1.

Trying the New and Unknown

The first factor of the accumulation mechanism is *trying the new and unknown,* which refers to the way in which new competences are generated through experimentation, problem solving, and facing new challenges. Dealing with new challenges forces people to try new technology, test different methods, and collaborate with others to solve problems. All these activities will generate new competences.

One young engineer stated, "I really like new development, to look forward to the next barrier, and toward new technology." In this case, the focus on technological knowledge also improves other dimensions of the competence concept. The ability to manage complexity and to learn improves when working on solving complex problems by linking different technologies in teams. This can be observed when watching younger engineers forced to use their creativity and search for new and different ways to solve problems.

Learning by Working

Learning by working refers to the way in which competences develop by applying knowledge and experience to real working tasks. Moreover, it concerns seeking information and practicing what one knows in daily work and, in this way, learning by working in terms of being creative, learning from mistakes as well as from successes, and having the opportunity to "dig down into tasks to solve issues," as described by a project team member. This factor relates to active learning, in which we use information and our previous knowledge and, consequently, not only gain new knowledge but also develop the other dimensions of the competence lemon—for example, developing social skills when solving problems in a group. Both group learning and sharing can be seen as part of this factor.

Participation and Sharing

The next factor is *participation and sharing,* which refers to learning by sharing information with team members, participating in different activities, and sharing knowledge. Examples can be sharing knowledge with others, establishing cross-functional networks, asking questions, or establishing a culture in which people share their learning with others in the organization. Both the individuals sharing their knowledge and experience and the individuals receiving them gain new competences. People who share their knowledge and experience often reflect on what is being shared and will accordingly gain more competences. This factor is related to both the sharing and the reflection factor when considering the factors that generate new competences.

Absorbing External Competences

The factor *absorbing external competences* describes how an organization can learn from external sources: its absorptive capacity. In many cases, a project is staffed with both internal and external competences, of which the external competences can be suppliers, an external workforce to fill gaps, or subject matter experts. They can also mean working closely with customers and other kinds of stakeholders. By actively learning from the external sources, new knowledge can be transferred to the organization. A project team member expressed this in an interview as, "Bringing in a consultant who is a specialist within the subject matter area adds competence to the organization, especially when the consultant brings experience from other clients."

However, a contradictory situation can arise between the parent organization and the project goals when the parent organization works actively to absorb external knowledge while the project focuses on project deliveries. As a project manager concisely expressed it, "The project has no responsibility that the competences stay within the company."

Cross-Functional Collaboration

The next factor in the accumulation mechanism is *cross-functional collaboration,* which describes a heterogeneous environment in which people with different competences work together toward the same goal. Many organizations have "silo thinking," meaning that people do not collaborate with others outside their function. A functional manager explained his view of this as follows: "The low degree of cross-functional collaboration and knowledge sharing is the effect and the disadvantage of working in a functional matrix." It is more demanding for a leader to manage a heterogeneous environment with people from different disciplines, cultures, previous competences, or subject matter areas. The focus for the leader in managing a heterogeneous team is to make the team share knowledge, experience, and ideas and to establish a team culture of trust and respect for each other's competences. As discussed in Chapter 1, a heterogeneous environment can have a positive impact on the generation of new competences.

Group Learning

The factor *group learning* in the accumulation mechanism refers to people learning in groups through sharing, feedback, reflection, and discussion. By working together, people learn from each other. It is important to establish an environment in which working together is stimulated. As mentioned earlier, specialists have a tendency not to participate in group learning, for which reason it is important to involve them in group learning as well to improve the work efficiency. This factor is directly related to the factors that have an impact on the generation of new competences, as discussed in the previous chapter.

Attitude and Motivation

Another factor in building up the accumulation mechanism is *attitude and motivation,* which refers to motivated and engaged people who are positive about working with problems and challenges and, accordingly, develop more competences. Being curious, having an interest in work, and being motivated increase the ability to learn and generate new competences. This area is a personal characteristic that facilitates the factors trying the new and unknown and learning by working. Having the right attitude toward work will make one more motivated and enable one to acquire new knowledge based on curiosity and willingness to learn.

Competence Development

The traditional view of learning is the factor *competence development,* which refers to formal competence development planning, including activities such as formal training, on-the-job training, and so on. Competence development planning is, in many cases, part of the organization's performance management

Table 2.2 Factors Constituting the Accumulation Mechanism

Factor	Description
Trying the new and unknown	How new competence is generated through experimentation, problem solving, and facing new challenges.
Learning by working	How competence develops by applying knowledge and experience to real working tasks. Moreover, it is about seeking information and practicing what one knows in daily work.
Participation and sharing	How learning occurs in teams, with members participating and sharing knowledge.
Absorbing external competence	How the organization and the projects can learn from external sources: the absorptive capacity.
Cross-functional collaboration	The process of learning in a heterogeneous environment in which people with different competences work together towards the same goal.
Group learning	When people learn in a group by sharing, feedback, reflection, and discussions.
Attitude and motivation	Having motivated and engaged people who are positive about working with problems and challenges to develop more competences.
Competence development	How formal competence development planning occurs.

process, in which activities for developing individuals' competences are planned in relation to the organizational goals.

The factors constituting the accumulation mechanism are summarized in Table 2.2. The factors in Table 2.2 are not weighted, although the context and the culture exert an impact on the importance of the different factors and on the way in which they are implemented and managed in the organization. For example, the organizational culture has a major impact on the factor group learning though trust, and stimulation to share knowledge is important for learning in a group.

Having described the accumulation mechanism, we continue to the next mechanism in the competence loop—*assimilation*.

2.5.3 Assimilation

The *assimilation* mechanism outlines the way in which new competences are assessed, understood, and interpreted.

Interpreting New Competences

The first factor in the assimilation mechanism is *interpreting new competences.* This factor refers to an informal organizational process that interprets individuals' competences on a frequent basis.

The nature of this factor is that leaders and colleagues understand individuals' competences, especially the new competences that an individual has acquired. There is no common method for this task; instead, the different functional managers interviewed have implemented their own ways of following up on performance and what people have learned. One functional manager uses what he calls "job chat," which is an informal but planned meeting with each employee every second week. The meeting has no agenda but focuses on how the daily work is progressing, how the team is working, or whether there are any problems or obstacles to discuss. The "job chat" is combined with being present at various project meetings and in daily work; these activities give the functional manager the confidence that he or she is capturing the employees' competences and how they are evolving.

Different from traditional performance reviews, various aspects of the employees' performance are considered, such as how they act toward each other, who takes responsibility, and who is willing to provide help and support. What these reviews capture are mainly the personal and social capabilities in the competence concept. One functional manager interviewed explained what he was looking for as follows: "I am not solely looking at the result and performance; I am also looking at how they communicate the result and how they interact with each other." This could be considered a more agile performance measurement method than traditional performance management processes.

In the project context, employees perceive in general that project managers are closer to the actual work than functional managers, and, in many cases, they also perceive that functional managers do not know their level of competence. A project team member stated, "My line manager does not know what I have learned. The project manager knows more, but in general it is the systems architect who knows my competence." What the employees mainly mean when referring to competences is their technical knowledge. To overcome this situation, functional managers can combine the agile performance measurement practices with frequent meetings with project managers and subject matter experts to capture both the project status and the employees' performance and, in this way, achieve a better understanding of the co-workers' competences.

Performance Measurement

The traditional way of measuring performance and competences is described in the factor *performance measurement,* which is the formal process by which managers assess individuals' performance based on goals set yearly. This is undertaken in a yearly performance appraisal, which in many cases is followed up on a six-month basis. In a knowledge-intensive environment, in which people are involved in different projects or other competence-generating activities, following up on competences on a yearly basis can be inefficient, because changes happen faster than the process can manage. The combination of interpreting new competences and performance measurement will be more efficient, especially if the goal-setting activities are considered based on fast-moving conditions.

Project Manager Feedback

As described above, in a project-intensive environment, people perceive that project managers know more about the team members' competences than functional managers. The factor *project manager feedback* refers to the role of the project manager in following up on the project team members' competence levels and how their competences are evolving. The pitfall is that the meetings between the functional manager and the project manager may focus only on the project performance rather than on what the employees have learned. A good example is mentioned by a functional manager who explained his way of working in this area: "I follow up with the project managers every week for about one hour. We are talking about what is working and not and the progress of the project. In those meetings, I also try to catch the employees' performance."

Poor Performance

In general, functional managers are more aware of the poor performance of employees than of performance that meets expectations. The next factor in the assimilation mechanism is *poor performance*. Performance is deemed to be "poor" if the functional manager notices that an employee is not performing in line with expectations. As one project manager put it, "We know when people do not know."

Poor performance could be visible—for instance, if a project team member always chooses the easiest tasks, delivers poor quality, or is always late in delivering. There is a tendency for project managers and functional managers to look at poor performance rather than at good or very good performance.

Poor performance can arise for different reasons—for example, not having adequate knowledge within the area or scoring low in other parts of the dimensions of the competence lemon—but the organizational culture can also affect people's performance. The important point is to analyze poor performance using all the dimensions of the competence lemon, and to undertake different gap-closing activities to improve the situation.

Learning from Projects

Because projects can be considered as competence development arenas, we need to learn from them. The factor *learning from projects* refers to the way in which new competences developed in projects are understood and transferred to other projects or to the organizational knowledge base.

Many project management processes involve lessons learned at the end of the project. Mostly these lessons learned focus on what went well and poorly in the project and whether the project goals were met. What is missing from traditional lessons learned is the competence perspective: "What have we learned in terms of new competences in the project?" Another consequence of identifying the lessons learned at the end of the project is that, in general, it takes place very late. It is better to follow up new competences continuously and determine how they can be used in the current project as well as in other organizational activities. Agile project management methods such as Scrum include retrospective meetings after every iteration, focusing on the process and not on the kinds of competences that the team members have attained. A similar retrospective activity focusing on learning from a competence perspective can improve the understanding of newly generated competences.

Measuring External Competence

In practically all cases, the external workforce is excluded from the formal process of following up on performance. The factor *measuring external competence* outlines the way in which the different people in the organization measure the competence level of the external workforce. The external workforce is used to fill gaps or as subject matter experts in many projects. Sometimes the same resources are used in several projects or over a long time period. Excluding the external workforce from performance measurement can prevent the external competences from being used optimally, which in turn can have an impact on the project outcome.

The functional managers interviewed gave different views on the ways in which they consider the external workforce. Some of the functional managers try to follow up on the external workforce's achievements by holding one-to-one meetings with its members as well. Other functional managers have a different view; as one functional manager expressed it, "I have no time for consultants. I expect that they deliver because they are brought in for doing a specific task. Their employer is responsible for their development."

However, if an organization uses an external workforce, especially to a large extent, and uses the same resources for longer terms, the competences of its members have to be followed up to gain optimal use of them. Inclusion will also lead to better attitudes and motivation, which are important factors for developing new competences.

Competence-Sharing Arena

The next factor in the assimilation mechanism is the *competence-sharing arena,* which involves people gaining the ability to share new competences with others within the organization. An interviewed project manager stated, "We are using different ways to share what we do. We use wikis, post newsletters, share tips and tricks, and other ways."

One method is to establish competence groups and other forums to share cross-functional knowledge. To establish well-working forums, it is important to focus on areas in which people feel that they can share and gain competences. One example is project manager forums in which project managers can support each other. Here the importance of a functional organizational culture that allows people to share their knowledge and experience is evident.

Process and Documentation

The last factor in the assimilation mechanism is *process and documentation,* which refers to the process or activity of documenting solutions, services, and products in such a way that others understand how the solutions were reached. Documenting solutions, services, and products is one way of making individuals' competences organizational. It is also important to document why a solution has been implemented in a specific way. The process is another way to work with organizational learning, because the process unifies the method of working with clear steps and documentation rules. Standardization in documentation and processes leads to improved knowledge sharing, because solutions and decisions are described in such a way that the reader can understand how the problems were solved and the basis on which different decisions were made.

The factors constituting the assimilation mechanism are summarized in Table 2.3. The factors in Table 2.3 are not weighted, although the context and the culture have an impact on the importance of the different factors and the way in which they are implemented and managed in the organization. For example, the context has a major impact on the factor performance measurement, because organizations implement different processes to measure performance. Another example is interpreting new competence, because the organizational culture can influence the communication between managers and co-workers.

Table 2.3 Factors Constituting the Assimilation Mechanism

Factor	Description
Interpreting new competences	This factor refers to an informal social process that interprets individuals' competences on a frequent basis.
Performance measurement	The formal process whereby managers assess individuals' performance through performance appraisals and similar activities.
Project manager feedback	The role of the project manager in terms of following up on the project team members' competence levels and determining how their competence is evolving.
Poor performance	How poor performance (performance that is not in line with the expectations) is measured in the organization rather than acceptable, good, or excellent performance.
Learning from projects	How new competences developed in projects are understood and transferred to other projects or to the organizational knowledge base.
Measuring external competences	How the different people in the organization measure the competence level of the external workforce.
Competence-sharing arena	How people share new competences with others within the organization so that others understand them—for example, through spontaneous talks, networks, or social media used in the organization.
Process and documentation	The process or activity of documenting solutions, services, and products in such a way that others understand how the solutions were determined.

Having described the assimilation mechanism, we proceed to the next mechanism in the competence loop—transformation.

2.5.4 Transformation

The last mechanism of the competence loop is the *transformation mechanism*, which is described as the way in which new and existing competences are combined, core competences are reconfigured, and competence gaps are identified.

Promoting Internal Mobility

The first factor in the transformation mechanism is *promoting internal mobility*, which represents the planning and process by means of which people can, as part of succession planning, move to new positions within the organization in which their competences can be better employed.

The view of a project as a competence development arena produces the need for a way to plan and promote internal mobility whereby new competences are assessed as part of the assimilation mechanism. The new position can also entail the resource being allocated to a new project in which their newly acquired competences are needed. Promoting internal mobility is an important part of organizational learning. If the newly generated competences are not used in a satisfactory way, there is a risk of co-workers either becoming demotivated or leaving.

Competence Mapping

The next factor of the transformation mechanism is *competence mapping,* which refers to the way in which competences are categorized and mapped. The basis for the categorization is the identified core competences of the organization. Those competences need to be evaluated specifically. In addition to the core competences, other competences need to be identified and categorized to facilitate their utilization. The important issue in competence mapping is to consider all the dimensions of the competence lemon.

Competence Transfer to the Organization

Another factor in the transformation mechanism is *competence transfer to the organization.* This factor refers to the actual transfer of new competences from projects or activities to the parent organization so that they can be used further. Competences can be transferred either as codified documented knowledge, by sharing among people, or by making the resources available for new assignments. "The knowledge is in the people's head" was an interviewed project manager's simple explanation of where knowledge is stored. People within the team share knowledge and thus preserve the knowledge in the team.

Identifying Core Competences

Core competences were described above as the capabilities that are critical to a business's achievement of a competitive advantage. *Identifying core competences* describes the way in which an organization's core competences are identified, assessed, and reconfigured. Core competences also need to have time dimensions—short, medium, and long term. The core competences needed in the next 12 months are not necessarily the same as those needed in the next three years.

Managing the Need for Competences

An important task for organizational leaders is to manage competence gaps. The factor *managing the need for competences* relates to the way in which the competence gap is managed by competence development planning, recruitment, sourcing strategies, and organization. Organizational strategies and goals are input to the project portfolio, which decides which competences are needed to realize the projects. The gap between the need and the availability can be managed by developing people, recruiting people with the right competences, using an external workforce, or reorganizing. The competence mapping activities can be undertaken on a high organizational level as well as in a smaller part of the organization, such as a department.

Identifying Internal/External Competences

The last factor in the transformation mechanism is *identifying internal/external competences*. This factor describes which competences at a strategic level will be internal to the organization and which competences will be sourced externally. Normally, an organization tries to keep its core competences internally and

Table 2.4 Factors Constituting the Transformation Mechanism

Factor	Description
Promoting internal mobility	Planning and processes allowing people to move to new positions within the organization in which their competence can be better employed.
Competence mapping	How competences are categorized and mapped.
Competence transfer to the organization	The transfer of new competences from projects or activities to the parent organization so that they can be used further.
Identifying core competences	How the organization's core competences are identified, assessed, and reconfigured.
Managing the need for competences	How the competence gap is managed by means of competence development plans, recruitment, sourcing strategies, and organization.
Identifying internal/external competences	The identification of the competences at the strategic level that will be internal to the organization and those that will be sourced externally.

use an external workforce or suppliers for activities that are outside the core business. The important point for an organization is to identify which kinds of competences, and how much of them, the organization should keep internally and in which areas gaps can be filled with external competences. Sometimes external expertise can also be used in core activities when needed.

The factors constituting the transformation mechanism are summarized in Table 2.4. The factors in Table 2.4 are not weighted, although the context and the culture have an impact on the importance of the different factors and the way in which they are implemented and managed in the organization. For example, the context has a major impact on the factor promoting internal mobility, as different organizations have different processes for managing internal mobility and promotion.

2.5.5 Summary of the Factors Constituting the Mechanisms in the Competence Loop

Figure 2.4 summarizes the factors constituting the four mechanisms in the competence loop.

The factors in the competence loop can be seen as processes that are either social or organizational. An organizational process is a defined process describing a set of activities and tasks linked to an organizational goal. A social process, on the other hand, is based on social interactions in which individuals

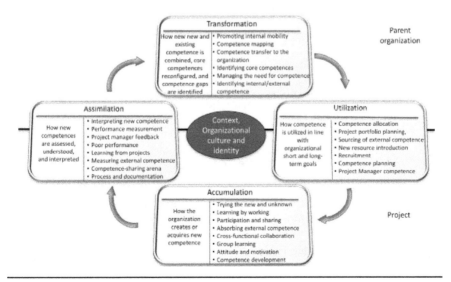

Figure 2.4   The factors constituting the four mechanisms in the competence loop.

Table 2.5 Factors Classified as Organizational or Social Processes

Mechanism	Factor	Organizational/ Social Dimension
Utilization	Competence allocation	Organizational
	Project portfolio planning	Organizational
	New resource introduction	Organizational
	Recruitment	Organizational
	Competence planning	Organizational
	Sourcing of external competences	Organizational
	Project manager competences	Organizational and social
Accumulation	Trying the new and unknown	Social
	Learning by working	Social
	Participation and sharing	Social
	Absorbing external competences	Organizational
	Cross-functional collaboration	Social
	Group learning	Social
	Attitude and motivation	Social
	Competence development	Organizational
Assimilation	Interpreting new competences	Social
	Performance measurement	Organizational
	Project manager feedback	Organizational
	Poor performance	Social
	Learning from projects	Organizational
	Measuring external competences	Organizational
	Competence-sharing arena	Social and organizational
	Process and documentation	Social
Transformation	Promoting internal mobility	Organizational
	Competence mapping	Organizational
	Competence transfer to the organization	Organizational
	Identifying core competences	Organizational
	Managing the need for competences	Organizational
	Identifying internal/external competences	Organizational

and groups interact and establish social relationships. Table 2.5 summarizes the factors in organizational and social processes. The transformation and utilization mechanisms are built on organizational processes, while the accumulation and assimilation mechanisms combine organizational and social processes. The competence loop links organizational and social processes to a common model for efficient competence management.

Looking at the different mechanisms, we can see that utilization is built on organizational processes. The utilization of competences is to a large extent process driven. People who are responsible for the project portfolio plan the different projects based on strategic priorities and assign competences to the different projects. Competence gaps are filled by recruitment or by external competences in terms of suppliers, contracting partners, or an external workforce, such as contractors or consultants.

All of those activities are considered as organizational processes. However, project manager competences can be considered as both an organizational and a social process. There is a need for an organizational process that assesses and maps the project manager's competence level, preferably based on the competence lemon, as a basis for the selection of the right project manager for the specific project. On the other hand, there is also a social process in which the manager who is responsible for the project and the project manager agree that the project manager will take on the project.

The accumulation mechanism is mainly based on social processes, because learning is, to a large extent, a social process. The factor absorbing external competences can be seen as an organizational process, because an organization's absorptive capacity is crucial for a sustainable competitive advantage. Within a project, external resources and suppliers are present, and learning from them is part of group learning, participation, sharing, and so on. Furthermore, the factor cross-functional collaboration should be considered as a social process, even though it has some organizational process elements. Organizational leaders need to have a way to promote and support cross-functional collaboration, because learning is greater in a heterogeneous environment than in a homogeneous one when it is working well and all the members are participating in the achievement of a common goal.

The assimilation mechanism is a mix of social and organizational processes. The formal performance measurement process can be combined with the interpretation of new competences—a social process—on a frequent basis. A more agile and continuous way of approaching performance measurement could be considered as a process of combining an organizational process using formal performance appraisals with a social process in which the manager understands the co-worker's performance and what the co-worker has learned. In addition, establishing a culture of sharing will increase colleagues' possibility of understanding and interpreting people's competences.

The factors measuring external competences—establishing formal project manager feedback and learning from the project—should be established as organizational processes to achieve an efficient follow-up of the new competences developed. Focusing on poor performance rather than good, great, or excellent performance is a social phenomenon that could complement performance measurement and the interpretation of new competences. However, we should mention that paying attention only to poor performance can lead to demotivation in the team, even if it is important both to identify performance that is not in line with the expectations and to try to manage this situation.

The competence-sharing arena is based on both social and organizational processes. The organizational leaders can implement a process whereby people share competences by holding competence-sharing events, such as innovation days or tech talks. They can also establish subject matter networks to align and share competences. The social process elements are people's active sharing of competences in a natural way—for example, through social media, spontaneous talks, and so on. Here the organizational culture is important in allowing people to share.

The last mechanism, transformation, is built on organizational processes. In an organization, core competences need to be identified and understood. Because core competences are not static, they also have to be maintained and developed, preferably linked to strategy work. Failing to identify core competences entails a great risk that the organization will end up with an imbalance in competences. Too many of some competences or a lack of others results in a competence gap. One senior manager described a situation in which identified core competences were lacking in the following way: "We have competences in some areas that we are not using and competence gaps in other areas. But we have the competences we have, and we have several challenges for the future. There is a gap between what we have and what we need, and we do not currently address the gap."

To be able to manage the need for competences, the competences need to be categorized and mapped. Based on core competences, the organizational leaders can identify which competences should be internal and which should be sourced externally. Without identifying the core or key competences, it is difficult to be efficient in competence mapping and deciding which competences should be internal and which should be sourced externally.

An important part of organizational learning is promoting internal mobility, so people can utilize their competences where they can be better employed, which also contributes to more highly motivated co-workers. Allowing people to move to new positions within the organization is an organizational process.

2.6 Exploratory and Exploitative Learning Strategies

The competence loop framework also consists of two learning strategies, the *exploratory* and the *exploitative* learning strategy. The exploratory learning strategy involves new competences being generated and integrated into the organization's competence base by working in projects or other value-creation activities. The exploratory learning strategy takes place in the accumulation and assimilation of competences. The exploitative learning strategy, on the other hand, entails newly generated competences being combined with existing ones and utilized in new projects or other value-creation activities and takes place in the transformation and utilization of competences. The two learning strategies need to be balanced to achieve a sustainable competitive advantage.

Returning to innovation, which is among the dynamic capabilities to achieve a competitive advantage, we can consider the capacity to innovate as a learning process (Oltra and Vivas-López, 2013) in which the most relevant feature is the uniqueness of knowledge (Cabello-Medina et al., 2011). Innovative capabilities can be *exploitative,* which means the reinforcement of existing knowledge, or *explorative,* meaning the transformation of existing knowledge into radical innovations (Biedenbach and Müller, 2012; Subramaniam and Youndt, 2005; Tamayo-Torres, Ruiz-Moreno, and Verdú, 2010).

In their study of innovative capabilities, Lisboa, Skarmeas, and Lages (2011) show that exploitative innovative capabilities have an impact on a firm's current performance and on its exploratory innovative capabilities, whereas exploratory innovative capabilities have an impact on a firm's future performance. This can be compared with the competence loop, in which the generation of new competences can be seen as exploratory learning used in the future while the transformation and utilization of competences are exploitative learning connected to short-term activities, such as starting new projects.

Because exploratory and exploitative innovative capabilities have different time horizons, an organization needs to balance exploitative and explorative innovation to achieve a sustainable competitive advantage (Lin et al., 2013; March, 1991).

Looking at exploratory learning, we can identify two parts: (1) the part in which people develop new competences, and (2) the part in which the new competences are interpreted and understood by the other people in the organization. Those two parts are symbolized by the accumulation and assimilation mechanisms in the competence loop.

A strong and positive culture—one in which people can learn from performing knowledge-intensive work, sharing knowledge, having time to innovate, having a positive attitude toward work, and being able to experiment—facilitates the exploration of new competences. Furthermore, organizational leaders interpret new competences by being close to the people working in the organization, unlike traditional performance management activities. This gives rise to a more agile view of performance management, which can be called *agile performance management,* in which leaders and managers continuously follow up employees' performance instead of, or at least as a complement to, holding yearly appraisals. Agile performance management will be further described in Chapter 6.

We can see two parts of exploitative learning: (1) transforming new competences into the organizational competence base, and (2) utilizing competences in new projects or other value-creation activities.

In the competence loop, those two parts are represented by the transformation and utilization mechanisms. When exploiting competences in an efficient way, an organization needs to understand which competences are available and utilize them efficiently in a project portfolio in which the organizational leaders have the ability to take decisions based on accurate information. Accordingly, the organization can adapt quickly to fast-changing market conditions, demonstrating its adaptive capability, as described by Biedenbach and Müller (2012) and Wang and Ahmed (2007).

By using exploratory and exploitative learning strategies, an organization can balance the generation and utilization of new competences to manage its competences efficiently in line with the organizational goals.

Having outlined the way in which competences can be managed effectively through the competence loop, and by using different learning strategies, the next chapter will investigate how three studied organizations use the competence lemon and the competence loop practically.

Chapter 3

Projects as Learning Arenas

If you feel trust in the team, you feel confidence to share!

— An interviewed project manager

3.1 Knowledge-Intensive, Project-Intensive Organizations

The nature of a project makes it an area in which new competence evolves. In general, projects aim to move something from point A to point B with different levels of uncertainties and unknowns. Knowledge-intensive organizations tend to organize a significant part of their work in projects, and for that reason we will look at competence management from a knowledge-intensive, project-intensive perspective, starting with a view of the organizational context, followed by different types of organizations' perspectives.

In brief, organizations are often structured in different units, such as business units, functional units, or geographical units (Turner, 2014). Those structures are designed to achieve efficiency and effectiveness. Based on the size, complexity, and characteristics of the tasks, we can find either a simple design that is characterized by flexible relationships and almost no hierarchy, or a functional design with departments based on specialization, logical similarity of the work, related or interdependent work tasks, and common goals. Examples include manufacturing, finance, R&D, and marketing departments. Each of these departments in a company can also be considered and studied as an organization itself (Hatch and Cunliffe, 2006). This view of an organization is adopted in the present book, meaning that the R&D department in a company, for instance, can be considered an organization. In this way, competence management can be looked at from a *company* perspective but also from a *departmental* perspective.

In organizations conducting projects, the project context can be seen as different systems with projects as "organizations within organizations," wherein the projects have structures different from the parent organization (Shenhar, 2001, 395). We call this kind of organization a *matrix organization,* in which the project is one dimension and the functional organization another. These different structures could lead to competing contextual demands in various levels of the organization (Bresman and Zellmer-Bruhn, 2013), and sometimes tensions between the project manager and the functional manager can be observed (Medina, Müller, and Bredillet, 2011).

In 1992, Starbuck labeled an organization in which knowledge has more importance than other inputs and human capital dominates the knowledge-intensive organization. Furthermore, to distinguish *knowledge* intensity from *information* intensity, he argued that knowledge is a stock of expertise and not a flow of information, and the organization should be based on valuable expertise. In a review of Starbuck's (1992) work, Kärreman (2010) elaborated on this theme, emphasizing that all organizations are to some extent built on knowledge, but that a knowledge-intensive firm draws on rare, specific, and abstruse knowledge. Moreover, a knowledge-intensive organization also tends to be ambiguity intensive, in the sense that this kind of organization works with a higher degree of uncertainty (Alvesson, 2011). Thus we can conclude that almost every organization today builds on a certain level of knowledge intensity.

The knowledge-intensive economy is growing (Sinha and Van de Ven, 2005), and successful companies will be those that manage their knowledge development and consider what knowledge means in their organization (Von Krogh and Roos, 1996). Understanding the concept of knowledge simplifies capturing, retaining, combining, connecting, and sustaining knowledge. In this economy, knowledge intensity can be viewed from different perspectives: those of work, workers, and organizations (Swart and Kinnie, 2003). The organizational and work perspectives refer to organizations in which most work is based on intellectual capacity and performed by a qualified and educated workforce that delivers qualified services and/or products (Alvesson, 2000; Swart and Kinnie, 2003). Furthermore, Alvesson (2000) mentions different kinds of knowledge-intensive organizations such as R&D, consultancy, etc., whereas organizations such as manufacturing firms are considered to be less knowledge intensive. Based on this argument, we can conclude that a company can have different levels of knowledge intensity in different parts of the organization—for instance, where parts of the organization, such as R&D, are more knowledge intensive than others, such as manufacturing.

The resources in knowledge-intensive organizations are in most cases referred to as *intellectual material* or *human capital,* where knowledge is of higher importance than other inputs (Swart and Kinnie, 2003). The authors also discuss the difference between creative jobs (for instance, advertising), standardized jobs (such as working with handicraft), and complex problem solving. In addition, they emphasize that creative and standardized jobs are less knowledge intensive than complex problem solving because of the application of the expertise of human capital. Strictly creative work does not involve complex problem solving, and a standardized job involves a high level of knowledge but is repetitive. Based on this reasoning, we can state that knowledge intensity is based on the application of human capital and on the change in knowledge through problem solving, experimentation, or learning. All work today has some kind of knowledge intensity, although we learn from doing and working.

3.1.1 Is There any Relationship Between Knowledge-Intensive Organizations and Project Intensity?

The importance of projects and project-based forms of organizing has grown in recent decades and will continue to increase (Söderlund, 2005; Whitley, 2006). In these organizations, projects are used to achieve operational and strategic goals, especially in knowledge-intensive, technological markets (Whitley, 2006). Söderlund and Bredin (2006) also make a connection between project orientation and knowledge intensiveness, arguing that organizations that are knowledge intensive often tend to be project oriented, where knowledge is a foundation for competence. Blindenbach-Driessen and Van den Ende (2010) continue on this theme, stating that knowledge intensity increases rapidly and is organized in project-based organizational (PBO) forms. Hobday (2000) draws the conclusion that PBOs are best suited for the management of products and services with high complexity in rapidly changing markets where it is important to combine knowledge, knowhow, and skills. Furthermore, Reich, Gemino, and Sauer (2012) mention the IT industry as a typical knowledge-intensive industry in which the core input material for IT projects is knowledge. Another link between project management and knowledge orientation is made by Akbar and Mandurah (2014), who state that project management is a process that is increasingly used to combine knowledge from different sources to add value to an organization. It can also be concluded that a project always has some level of "unknownness" or uncertainty. We can hereby see a close link between knowledge orientation and project organization.

3.1.2 What Is a Project-Intensive Organization?

There is wide research on PBOs in which, according to Lindqvist (2004), a PBO is an organization that carries out most of its activities in the form of projects, and the project dimension is stronger than the functional dimension. Hobday (2000) takes this definition further, stating that a PBO in its extreme form has no function at all and, therefore, no coordination across project lines. This kind of organization is called a "pure PBO" in this work.

Furthermore, Whitley (2006) categorizes project-based firms in four categories based on the singularity of outputs and goals, and separation and stability in work roles. The interesting point in Whitley's categories is that organizations with high separation and stability in work roles (e.g., advertising, crafts, and construction) tend to be the kinds of organizations that Swart and Kinnie (2003) argue are less knowledge intensive. What is significant for those types of PBOs is that not only do they rely on specialized work roles, but that they also deliver either single or incrementally repeatable outputs.

Arvidsson (2009) looks at different types of project-oriented organizations and discusses what he calls *projectified matrix organizations* (p. 98), where he distinguishes between project-based and project-oriented firms. In line with Lindqvist (2004), Arvidsson (2009) argues that PBOs are organizations in which a majority of the revenue and costs is associated with temporary structures and processes, whereas in project-oriented organizations, revenues are generated through products and services in the permanent functions, but a major share of costs are related to projects.

Keegan, Huemann, and Turner (2012) also bring in temporariness from a human relations perspective, stating that tensions can arise between the permanent line organization and the temporary project. However, in most cases, is a project a temporary organization, or is it just a way to carry out tasks within permanent organizational structures?

Winch (2014) challenges the approach to projects as temporary organizations, and argues that project organization is in most cases achieved through relatively permanent forms of organization. This argument means that projects in most organizations are executed in a permanent context—meaning a line organization that manages the development of products, services, solutions, or changes using projects as a working method.

There has been wide research conducted on pure PBOs but, as Keegan et al. (2012) emphasize, organizations with permanent matrix structures, which manage a considerable amount of work in projects, are far more common than pure PBOs.

Most project-intensive organizations combine a permanent line of organization with the performance of tasks in projects—in other words, matrix organizations.

Derived from Galbraith (1971), Larson and Gobeli (1987), Pinto and Rouhiainen (2001), and Turner (1999), and in line with Hobday (2000), Medina, Müller, and Bredillet (2011) describe three types of organizations related to projects—namely, functional organizations, pure project organizations, and matrix organizations. Depending on the nature of the matrix organization, the weight could be on either the function or the project, or be balanced between the two. This book adopts what Medina et al. (2011) call matrix organizations but calls them *project-intensive organizations,* in which there is a coexistence of a functional organization and projects, and where a considerable part of the organization's activities is conducted through projects. In these matrix organizations, the power can be with either the functional organization or the project, or balanced between the two.

3.1.3 Summary of Knowledge-Intensive, Project-Intensive Organizations

In summary, a knowledge-intensive and project-intensive organization can be defined as an organizational entity within a firm in which the knowledge intensity is built on intellectual capital and new competence evolves through problem-solving, experimentation, innovation, or learning, and in which a considerable number of the organization's activities are conducted as projects. Examples could be firms that develop technological products in an innovative R&D department and rely on operational manufacturing and logistics departments. However, as previously mentioned, in today's market, almost all organizations are to some extent knowledge intensive and use projects as working forms, including in operational entities such as manufacturing and logistics. However, the aim, size, time horizon, uncertainties, complexity, etc. differ from project to project. What is common for all projects, though, is that we learn from them and we need the proper competence to perform the activities and tasks in the project. Another feature of projects that contributes to generating new competence is that they inherently involve working with new challenges, problems, and tasks. The ability to solve problems or face a challenging task increases a person's competence.

As previously outlined, projects play a role in creating and developing new competences. One benefit of a project is that different competences are allocated to a project team, and the project team has to combine these competences to attain the project's goals. Cross-functional project teams have different benefits. Firstly, project team members can learn from each other when different team members have different competences. Secondly, working in cross-functional projects encourages people from different functional areas to get to know each other, which increases the ability to know "who knows what."

However, there are contradictory goals in achieving more effective competence utilization in an organization. Usually, the project manager is focused on the *project*'s goals and outcomes, not on the *organization*'s goals. Several interviewed managers emphasized the need for better communication between the line organization and project teams to ensure that long-term business goals in terms of competence were fulfilled.

One of the interviewed business managers expressed the situation in the following way: "The project manager looks at the project outcome, not at the organization's goals. As a business owner, my view is longer than the project, while the project manager only has the project view, which sometimes leads to different target images and tensions." Another interviewed manager took a knowledge transfer perspective and expressed it thus: "A good project should be formed in such a way that it does not just focus on the final product but also facilitates knowledge transfer."

Another important aspect is how to preserve competence added to a project by an external workforce. In many projects, the number of external team members or suppliers is high, and when the project is completed, they leave, taking their competences with them.

In a knowledge-intensive, project-intensive context, many different roles occur. The rest of this chapter will look at managing competence from the perspective of the different roles and in different types of organizations. Three different types of organizations were studied to gain different perspectives on managing project competence.

- The first is a public sector organization that, to some extent, relies on external competence and for which the whole operational business is carried out by external parties.
- The second is an R&D department in a fast-growing company acting in a high-tech market. This organization uses internal resources to develop new products.
- The third and last is an IT organization in a declining company acting in the consumer electronics market. This organization uses an external workforce to a large extent—both contractors and outsourced IT development.

Based on these three organizational perspectives, we can see how the factors in the competence loop need to be used in different ways and that different weights are put on different activities.

However, we will also see that there are many similarities, and the bases are the same independent of the type of organization or the context. Moreover, we will see that the context and the organizational culture will impact the management and application of competence.

3.2 Case 1: The Public Sector Organization

When I really learn something new is when we work together towards the same goal and discuss how to solve the problems in the project. With the people in my team, we can share ideas and discuss which solutions are the best.

— Project team member

Our first case study is of a public sector organization in Sweden. This is just an example of a public sector and can be representative of the public sector as a whole or of private organizations acting in a comparable context. The case is studied from the perspective of how the organization manages the different mechanisms in the competence loop. The context is described first, followed by how the organization is using the different factors in the competence loop. *[All references to the competence lemon can be found in Chapter 1, and references to the competence loop can be found in Chapter 2.]*

The studied organization has around 300 employees who mainly work on planning and follow-up, new products and services, future strategies within the area, and information technology (IT), which makes it a knowledge-intensive organization. Many new products and other activities are carried out using projects as the working method. Private companies on contracts provide services to the community.

As a public sector organization in Sweden, it has to follow the recruitment specifications strictly, otherwise the applicants can appeal against a recruitment decision. The effect is that the organization has to be careful about how the requirements are formulated before publishing an official employment advertisement. Another implication that affects recruitment is that employees in public authorities normally have lower salaries than those in private companies, making it difficult to attract competent people, which in many cases can lead to a dependency on external competences.

In this organization, IT project management was found to be an area in which people perceived a low maturity level. The project concept is misused to some extent: some activities are called projects although they are minor, whereas some activities should be carried out as projects. This issue, and the fact that the power lies with functional units, leads to a situation in which it is difficult to start, follow up, and finalize projects successfully.

There is a clear difference between starting a project and correctly initiating it. Many projects are started without defined business benefits or linkage to the organizational strategy. Resources are not properly allocated; rather, people are happy to help with the project but without making a commitment to it. In the absence of a working project methodology or project management culture, there are no clear requirements concerning what is required to implement a project. Furthermore, there are no clear definitions of the role of the project manager, the steering committee, or other roles important in relation to project management. The result is that the project scope, roles, and responsibilities are not clear to the people in the organization, leading to deviations between stakeholders' expectations and project outcomes. One of the project managers described it as follows: "The definition of a project is unclear: it can be anything from what a person should do as a regular working task to large projects lasting several years."

3.2.1 Competence Utilization

The organization has grown from a small public sector organization to a rather large actor within their operation, but the organizational structures have not been established to manage this change. The combination of low project management maturity, the organizational structure, and the difficulty in attracting highly competent employees leads to challenges in reaching the authority's strategic goals. Moreover, the power in the organization lies with the functional departments, leading to functional silos, which has an impact on competence utilization and leads to a high degree of specialization and a low degree of sharing. One senior manager concisely described the situation thus: "We are too stuck in the functional silos!"

The management has an ambition to create an open, flat organization in which people can work together and participate. Furthermore, the organization has grown from a small organization to one of considerable size while attempting to retain the organizational culture of the smaller organization. In addition, the main objectives have changed from being a provider of operational services to having an increasing focus on sales, customer value, and acting more commercially. Changes in leadership style and recruitment have supported this change in business orientation and have impacted the organizational culture, from having a small-company feeling to being more direct and based on power.

Analysis of the organizational culture showed that it deviates from the version described in the business plan. The business plan describes its values as "welcoming," "driving," and "engaged," but analysis showed that the culture was characterized by other values, such as the power structure, avoidance of decision making, and functional silo orientation. The discrepancy between the organizational culture being presented and the present behaviors has led to an unstructured way of working, in which it is easy to start up new activities and projects, but, because of the difficulty in making decisions, people do not take enough responsibility, and only a few projects are finished properly. People want to be involved in many activities, but there are few—the specialists—who work hard and deliver. As one of the project team members put it, "There is a culture that everyone shall be involved. Many are on board but only a few deliver."

As specified, the organizational culture leads to a tendency to focus on daily operational issues and avoid development and innovation. This approach is exemplified by the fact that there are no real processes for new demands or strategic projects; instead, there is a well-developed process for the maintenance of system solutions.

Looking at the utilization mechanism in the competence loop, the organization mainly struggles with the strategic process and immaturity in project management methods, which leads to difficulties in planning strategic projects and efficient utilization of resources and competence.

The organization suffers from not having a properly functioning strategic process. There is a high degree of focus on operational issues and what the long-term future customer behavior should be like. As one of the senior managers expressed it: "We are good at talking about the here and now, and about the long-term vision. But we do not talk much about how to reach the vision." This weak strategic process leads to a low degree of focus on strategic projects; instead, the focus is on planning important activities for specific functional departments. Here we can see the absence of project portfolio planning factors leading to a lack of focus. Project portfolio planning has an impact on the project outcome through the method of project initiation and control. Moreover, it has an impact on competence utilization, because project portfolio planning provides an overall picture of how competence is allocated to different projects, which facilitates the prioritization of competences between projects. The consequence is that there is a tendency to start projects without having control of the project scope, requirements, or competence needs. Another consequence is that the organization starts too many projects and fails to finish them.

There is a tendency to recruit specialists whose previous knowledge and experience are of the highest importance. Taking other dimensions of the competence lemon into consideration—such as personality and the values required for collaboration and teamwork—would enhance the social capital and improve organizational performance. Relying on a few specialists affects the organizational performance, as activities are dependent on the specialists performing their roles, leading to a few—the specialists—working hard and delivering while others have a lower performance. In many cases, the experts participate in different projects but only for a limited period of time. Examples of experts could be within different IT systems or IT infrastructure, but also within business development. The experts do not have a holistic view of the project scope or what the project aims to achieve. On the other hand, the experts are skilled in their subject area and can perform their tasks with speed and with high quality.

Another effect of having many experts is the impact on knowledge sharing. Many people perceive that knowledge sharing is more difficult because of the reliance on experts: (i) the experts do their part and do not participate in activities outside their areas of expertise; and (ii) there is a distance between different experts' subject areas and between the experts and the others in the project team.

3.2.2 Competence Accumulation

The next mechanism in the competence loop, accumulation, focuses on how new competences are generated and developed. Having no fear of making mistakes and testing and experimenting were emphasized as important learning enablers. However, the culture prevents a high degree of experimentation, because there is a fear of making mistakes, as concisely stated by a project manager: "Too many people are afraid of making mistakes." The fear of making mistakes is seen as a barrier to working with new solutions and services. Here we can see the value of an organizational culture built on trust and togetherness. In the functional teams, people in general felt safety, for which reason group learning worked well. The issues arose when cross-functional teams were established. The high degree of specialization hinders learning, because the specialists simply carry out their own tasks, and their competences are too discrete, which has an impact on sharing.

The organization relies to rather large extent on external parties such as suppliers and consultants and faces some issues concerning the preservation of the competence when the external party leaves. However, several actions have been taken to improve learning from suppliers, and that awareness, together with implemented actions, have improved the situation. Here we can see that organizations need to have a process to improve their absorptive capacity and learn from external parties, preserving knowledge within the organization.

What became clear in the studied organization was that the focus is on process development and documentation. Interviewees emphasized that it gave time for reflection on the evaluation of project activities and the way in which tasks were solved. Moreover, understanding and interpreting new information provides new insight that stimulates learning by reflection. The focus on process development and documentation had a positive impact on generating new competence, although knowledge could be shared in a structured way.

3.2.3 Competence Assimilation

The next mechanism in the competence loop, assimilation, outlines how new competences are assessed, understood, and interpreted. This mechanism is important for organizational learning, which is the dynamic process whereby knowledge moves from the individual to the group level, and from there to the organizational level and back again. Managers in the public organization used different ways to understand and interpret new competences generated in activities or projects. The organization used a formal performance appraisal process with a development conversation once per year and a follow-up talk after six months.

In addition to the formal performance appraisal process, functional managers used informal chats to follow up on employees' performance. Many people worked on various projects with people from other departments, and they worked with different project managers who had more insight into their performance and that worked more closely together. There was no systematic way to feed back an individual's performance to the functional manager, and here the project manager has an important role. In several cases, the project managers' focus is on "what people deliver in the project, not what they learn." The consequence is that managers miss some of the new competences people gain by working on a project. It was clear that the formal appraisal process could be developed to work more effectively and move on to a more agile performance measurement. In general, sharing of knowledge worked well in the organization, and colleagues and managers knew well an employee's level of competence.

As mentioned above, the organization has started to work in a more process-oriented manner, which has led to improved documentation, which in turn has led to improved knowledge sharing wherein solutions and decisions are described in a way that the reader can understand how the problems were solved and the basis for making different decisions. The process-oriented way of working has increased organizational learning and understanding of the competences that people develop.

3.2.4 Competence Transformation

The last mechanism in the competence loop, transformation, is described as a means of combining new and existing competences, reconfiguring core competences, and identifying competence gaps. In this mechanism, executives show a weakness in identifying the core competences—that is, the competences that define a firm's fundamental business. One example is that the organization is overly endowed with some competences and lacks others, resulting not only in the problem that important activities cannot be finished in time but also in low productivity. The current situation was explained by a senior manager: "We have the competences we have, and we have several challenges for the future. There is a gap between what we have and what we need, and we do not currently address the gap."

Another important factor for organizational learning and transformation of competence within the organization is promoting internal mobility, which represents actively planning the way in which people can move to new positions within the organization where their competences can be exploited to a greater extent. Low engagement in promoting internal mobility has resulted in people gaining a position and staying there. This leads to specialization, homogeneity, less interest in other areas, and competence not being used optimally. Although some good examples exist, in such cases it was the functional manager who noticed competent people and tried to support them by moving them to a better position.

This organization showed a weakness in transforming strategy into a strategic project portfolio meeting the organizational goals. The organization's core and key competences should be a result of analyzing the relationship between competences and the organization's purpose and business goals (Eden and Ackermann, 2010). In this particular case, the lack of a clear strategic process resulted in confusion about which core competences the organization needs. However, the transformation from a rather conservative organization to a more business-oriented service provider has started a process for identifying the future core competences, which also shows an awareness of the need for such identification.

3.2.5 Summary of Case 1

This case highlights that the organizational culture has a major impact on competence management. A well-functioning learning organization is based on an open, adaptive, and collaborative organizational culture (Liao et al., 2010). Furthermore, the organizational culture can have an impact on organizational performance (Flamholtz and Randle, 2011; Kotter and Heskett, 1992) and works as a bridge between different functional departments, especially in siloed organizations (Flamholtz and Randle, 2011).

The study of this case shows that many people avoid making decisions if they do not feel that they have the mandate to do so. Several interviewees suggested that people take on tasks but are not ready to take responsibility for completing them. The major impact is that there is a tendency to focus on operational issues instead of looking forward. Flamholtz and Randle (2011) call this organizational culture "the Hamlet syndrome." This syndrome has an obvious impact on making timely decisions: people try to avoid making any decisions and transfer the responsibility to someone else. Furthermore, this can affect the change and innovation dimensions through people becoming risk averse and trying to avoid making mistakes. Another aspect of this kind of organization is that there may be a blaming aspect, leading people to believe that good performance means avoiding making mistakes and hence also avoiding making decisions. The syndrome, as outlined by Flamholtz and Randle (2011), could explain the lack of strategic processes and goals in this specific organization. The positive side of this organization was its awareness of the need for change. Change was expressed by the process of identifying core competences and working on establishing business processes and improvement activities to meet future needs.

The strength in this organization was in the assimilation mechanism of the competence loop. Managers worked on understanding and interpreting the employees' competence levels, people shared knowledge in groups, and implementation of the process-oriented way of working—together with the documentation level—supported a common view of people's competences.

This case is an example of a public sector organization but also highlights the importance of context in the management of competence. Many of the findings are not specific to a public organization but can also apply to private companies. In the next case, we will look into a fast-growing organization based on research and development of high-tech products using internal resources.

3.3 Case 2: The Fast-Growing R&D Department

If you give them [the employees] freedom, they will
take on a higher degree of responsibility.

— Functional Manager

The next case to study is the research and development department in a fast-growing company acting in a volatile high-tech market. The company has a policy of recruiting highly productive potentials rather than working with different suppliers. The managers believe in a positive and functional organizational culture, encouraging working together, problem solving, and togetherness. As was the case of the public sector organization, this organization is studied from the perspective of how the organization manages the different mechanisms in the competence loop. Five sub-departments are compared to gain an overall view of the organization. The context is described first, followed by how the organization uses the different factors in the competence loop. *[All references to the competence lemon can be found in Chapter 1; references to the competence loop can be found in Chapter 2.]*

The company does not directly sell its products to end customers; rather, it acts via partners and distributors in a global market. At the time of the case study, the R&D department consisted of about 800 employees. The company works actively with new innovations, both exploratory innovations of new product types as well as exploitative innovation, continuously improving current products.

The philosophy of the company is that the R&D department should be located in the same place as the headquarters. The company has grown fast in the last couple of years, in revenue, profits, and in the number of employees. Based on this growth, the company has worked hard on the recruitment and integration of new employees in the organization.

The organizational culture is considered strong. The company has invested time in communicating and reinforcing the culture, ensuring that people understand it clearly and behave in ways consistent with the culture. The organizational culture can also be considered functional and positive, based on Flamholtz and Randle's (2011) description of culture as an asset to the organization. As an HR business partner put it: "The important values in the organization are passion, sharing, and responsibility." These three values were found in different forms in the study of the company. The organizational culture is based on the notion that people should help each other, cooperate, show respect, commit to decisions, and have fun together.

Furthermore, it is based on the idea that people should adopt a holistic view, act today, be able to make decisions, challenge themselves, push boundaries, and achieve big changes step by step. Finally, it is also based on an understanding that people should be innovative, transparent, honest, available to customers, responsive, and always consider new ways of working. The organizational culture was established by the founders, who focused intently on the culture and the importance of the core values, and in this way created the cultural DNA.

In the R&D organization, the project managers report to the functional managers in each area. The functional manager has more power than the project manager in terms of defining priorities, moving resources, and undertaking other activities that could have an impact on projects. The projects are mainly executed within the functional area, with low levels of cross-functionality; therefore, there is a tendency to build silos. In this respect, the R&D organization can be considered a functional matrix organization, in which the power is in the functional area and the project manager is primarily a coordinator. Especially when agile project methodologies such as Scrum are used, the project manager has little influence on project execution. Instead, decisions are made by product managers, functional managers, and the executing Scrum team. In addition, the different R&D departments have the freedom to choose project methodologies within certain frames.

Some of the departments use pure agile methodologies such as Scrum, while others have adapted more traditional waterfall methodologies. The R&D department uses a tollgate model for decisions, but it is not always followed. Furthermore, the agile project methodologies are implemented differently among departments. The different ways of managing projects affect the ability to manage cross-functional projects, because people are used to different methodologies. The freedom to choose a project methodology could be related to the organizational culture, which encourages people to think outside the box, be innovative, and try different ways of doing things, giving responsibility to the team.

3.3.1 Competence Utilization

The utilization mechanism in the competence loop concerns how competence is utilized in line with the strategic goals of the organization. In the current case, utilizing competence to reach the organizational goals is carried out through managing projects or other activities to develop new solutions or products, primarily using internal competence but also external competence when required, even if external competence is kept to a minimum. The roadmap controls what solutions and products are developed and when. Usually, a product is released two to four times per year with new functionality. The roadmap is initiated by the product management department and is originally based on the business plan. The time horizon of the roadmap is three years, but it is only developed at a detailed level for the forthcoming year. Twice a year, the roadmap is reviewed and updated by the product managers, and at the same time it is broken down into projects that could be considered a portfolio.

The R&D department works to large extent with agile product development methods, such as Scrum. The roadmap is the foundation for the product backlog (the list of all user stories for a product) in the different Scrum projects, and there is a continuous process of prioritizing in the backlog as a result of changing market demands.

Although the roadmap is fixed for six months, and the market demands change faster, the projects do not follow the roadmap 100 percent. Rather, the projects, the functional manager, and the product manager agree on the project scope and prioritization concerning the functionality to be implemented within the scope of the project. The release policy also leads to short projects, normally around three to six months, which facilitates project scoping.

The functional matrix organization, in which projects are executed in the functional departments, facilitates competence allocation to projects, although usually the same team continues to develop the next release of the product. Changes to the team are normally made through the introduction of new employees, for which the team is responsible. The organization struggles with establishing cross-functional projects because the functional silos are strong, with different departments using different development methodologies, and project management in general is very weak.

The recruitment process focuses on cultural fit rather than on previous knowledge and experience, even if those are considered important for the recruiting manager. The candidate has to pass an interview with senior management and the HR business partner to ascertain if he or she fits into the organizational culture. Instead of focusing on the competence lemon as a whole, the recruitment process focuses only on some parts, such as social and personal capacity and, to some extent, knowledge and experience.

3.3.2 Competence Accumulation

The next mechanism in the competence loop, accumulation, is related to how new competence is generated and developed in the organization. What is of particular note is that the social dimension seems to be more important when looking at which factors have an impact on generating new competence. The most important factors from this perspective are group learning and social context, while trying the new and unknown and learning by working are the most important organizational aspects. The level of collaborative work is high in this organization, explaining why social factors are highly valued. Working with Scrum teams that act in functional silos leads to homogeneous teams in which cross-functional collaboration is low. Project managers are appointed in the organization but they act more as coordinators or proxy product owners in the Scrum setup.

The culture also encourages time to learn and time to innovate. The different departments have different activities to support innovation; most of them allocate time for innovation days or similar activities. For example, one of the departments has a week set aside twice a year in which the employees are able to work with new ideas, experiment, and test new product ideas. Such activities encourage learning and the development of new competence. However, the importance of the different functional departments tends to result in the organization remaining in silos, making cross-functional collaboration more difficult. As a functional manager expressed it: "We do not share enough: the functional structures and borders make it difficult. Your functional organization is important, and when someone wants to move to another part of the organization or someone requests a resource, the first thought you have is that it might have an impact on your functional unit." The open culture improves knowledge sharing and learning in teams. Learning between teams is less visible in the organization, although people are encouraged to help each other and always respond to questions when there is an issue or problem to solve.

Another impact on group learning is individuals' social capability. The R&D department has employed many engineers with a tendency toward being introverted, which affects group learning. One project team member explained the negative impact in the following way: "There are a lot of introverts here: they sit in their rooms and look at their screens. It has a negative effect and is not good for knowledge sharing. You have to run around and ask if you want to know something."

The absorptive capacity in this organization, as described in Chapter 2, is low because of the low trust in external suppliers. There is a feeling that if you share information with external parties, they can use it for other purposes, and "we" are better than "they" are.

3.3.3 Competence Assimilation

The third mechanism in the competence loop, assimilation, is how organizational leaders and colleagues interpret and understand the competence a person has acquired. In the R&D department, this mechanism is supported by individual solutions, with different functional managers having different ways of following up on employees' performance. The projects are usually executed within functions, which facilitates the functional managers' following up on how employees' competence is evolving. The functional managers try to develop agile performance management practices, checking on each employee on a regular basis and being close to the daily business. However, the employees perceive that there is a greater distance between them and the functional manager than between the project manager or the technical lead, and that the functional manager knows less than the others in terms of their competence levels.

The employees are generally engineers and are thus focused on technological knowledge, whereas the functional managers focus on other dimensions in terms of how employees interact with others, their performance, and their level of responsibility. In this case, we can see that the functional managers try to use several of the dimensions of the competence lemon while evaluating co-workers' performance, but also that engineers in general value knowledge. One functional manager explained what he was looking for as: "I am not solely looking at the result and performance. I am also looking at how they communicate the result and how they interact with each other." This could be considered a more agile performance measurement.

3.3.4 Competence Transformation

In the fourth and last mechanism in the competence loop, transformation, people in the organization show a weakness that is due to the lack of understanding of core competences, which can be expressed as the competences that define a firm's fundamental business. The concept of core or key competences is crucial for all people acting in an organization to understand, especially for senior management, but also for people conducting the development of innovative products and services. Competence is important in this particular organization, and the career ladder, which defines an expert or senior expert, also considers the non-technological dimensions of competence, such as helping others to perform their jobs well. However, because the R&D department works in functional silos and documentation is not valued highly, knowledge is kept in people's heads.

Besides knowledge being kept in people's heads, the company also does not actively support internal mobility, with the consequence that people do not move to other positions. Those two aspects have a negative impact on organizational learning and on the transformation mechanism in the competence loop.

However, the product development department works actively with innovations, and innovation is important for future growth. The different R&D department managers both allocate time for working on innovations and also encourage people to use time for experimentation and trying other ways of solving problems. When the departments have innovation days or similar activities, people from the patent department participate to capture new product ideas and new innovations. In the innovative environment, the analysis shows a contradictory situation. A strong functional hierarchy negatively affects innovation across departmental borders. Several employees reported that when they mixed with people from other departments, the openness to innovation decreased. A project manager expressed it thus: "I have heard from my team members that they mix the groups too much, and that this is a barrier for people to be innovative."

Innovation in this organization is linked to the teams in the individual functional departments, which promotes incremental exploitative innovation but prevents radical exploratory innovation linking different technological areas together to create innovative solutions. The risk in this organization is that the focus will be too much on incremental innovation and radical innovation will be suppressed. The balance of exploratory and exploitative innovation is important to reach a sustainable competitive advantage (Lin, 2013) and is part of the learning strategies described in Chapter 2.

3.3.5 Summary of Case 2

The organizational culture in this organization is strong and clear. The founders realized the importance of a clear and strong organizational culture, and their values became the DNA of the organization. This is in line with Flamholtz and

Randle (2011), who argue that cultural DNA is generated when the personal and professional values of the founders define the organizational culture of the company. An organizational culture that is an asset for the company, clearly understood by the people in the organization who then behave in line with the organizational culture, can be considered strong and functional (Flamholtz and Randle, 2011). Almost every manager talked with or observed shares relatively common values and behaviors in a strong culture, and new employees are quick to adopt these values (Kotter and Heskett, 1992). In this organization, people talk about the culture and live it to a large extent. The culture was established when the company was much smaller, but the company has succeeded in maintaining it while growing. Furthermore, this organization mainly works with internal resources, for which reason there is a low focus on absorptive capacity. However, the risk of neglecting external knowledge is high and can lead to lower organizational learning in the long term, even if the organization is the market leader in its segment. Absorptive capacity—meaning learning from external sources such as suppliers, contractors, customers, and others—is fundamental for sustaining competitive advantage.

An interesting finding was that the team in general is of high importance, especially in the departments that operate according to the Scrum methodology and in which the teams show a high degree of self-organization. In the interviews, several interviewees expressed the importance of the team, which was viewed as more important than the individual team members. The analysis shows that team importance has an impact on job rotation and internal promotion. In the description of the fundamental interpersonal relations orientation (FIRO) model, Schutz (1958) defined the openness phase as one in which the team members feel trust in the group, are able to express both negative and positive feelings, are open to discussing most topics, and have a strong sense of loyalty to the group. Furthermore, the members feel secure, know that everyone is appreciated, and show great faith in each other.

Schutz (1958) described three phases that the team goes through—namely, inclusion, control, and openness. Between these phases are two intermediate phases—comfort and idyll. To reach the openness phase, the team has to go through all the other phases. Several of the teams showed strong feelings and loyalty for the team, which is in accordance with Schutz's (1958) description of the openness phase. This can be explained by two factors:

- Firstly, Scrum as a development method supports a self-managed team that works through retrospective meetings in which members' feelings and reflections are discussed. Furthermore, the team relies on everyone taking collective responsibility and working closely together to reach the organizational goals.

- Secondly, two of the core values strongly support working together and building loyalties. Because the organizational culture is strong and functional and the firm actively recruits people who will fit into the culture, the employees become loyal team members.

These two parameters could explain why teams are strong, exhibiting a high degree of loyalty that prevents people from leaving the team and moving on to new positions or to other teams. Another negative consequence of high team loyalty is that the team becomes too homogeneous, which will lead to inefficiency in the long term. A team needs external influences to increase learning and develop new competences.

Scrum as project methodology has a tendency to encourage small teams that do not change over time. The positive effect of this is that the team members learn each other's weaknesses and strengths, and it is easy to take on new tasks. The other side of the coin is that new perspectives not are taken on board, compared with teams that change team members more often—that is, learning from others with other experiences, ideas, and knowledge.

The strength in this organization was in the accumulation mechanism in the competence loop. The company established several activities to generate new competence encouraged and knowledge sharing by leaders.

This case was an example of a fast-growing research and development organization, but it also highlighted the importance of context in the management of competence. Many of the findings are not specific to this particular type of organization but can also be applicable to other types, especially in how the organizational culture impacts competence management. In the next case study, we will look into an IT organization in a declining organization which, to a large extent, depends on external competence in terms of contractors, consultants, and suppliers.

3.4 Case 3: The IT Organization in the Declining Company

Time is important, we have to deliver on time. Some consequences are that competence development has low priority, and we lack time to share our experiences and knowledge.

— Project team member

The last case we will study is the IT department in a declining company acting in the consumer electronics market. This is a global company, which is present on all continents. The IT department has a common function and acts toward all business and market units within the company; its responsibility within the company is to deliver IT solutions to the business organizations and ensure agreed maintenance and support for delivered solutions. Similar to other parts of the organization, the IT department has been forced to decrease its budget and number of employees. As part of this downsizing, the development and maintenance of business-critical systems has been outsourced to suppliers in India. *[All references to the competence lemon can be found in Chapter 1; references to the competence loop can be found in Chapter 2.]*

Today the IT department is a slimmed-down organization with solution managers, project managers, and experts that are either employed or external contractors but are considered as an internal workforce. All system development and maintenance is performed by outsourced parties. Project management is organized in a project management office (PMO), where all project managers are gathered. The project management methodology is based on traditional waterfall-oriented methodologies, and project management could in general be considered as mature. Three large-scale projects are compared to gain an overall view of the organization. First the context is described, followed by an explanation of how the organization is using the different factors in the competence loop.

The organizational culture is considered as weak, as evidenced when employees have difficulties in understanding, defining, and explaining the culture (Flamholtz and Randle, 2011). In this particular case, there are differences in how the various departments act, a low level of transparency in information, and silo thinking. Furthermore, there are tensions between offices in different countries and a low level of collaboration between different functions.

The high degree of external workforce utilization adds difficulty to establishing a common culture. In the projects, the project manager needs to establish a common culture to facilitate knowledge sharing between project members, as outlined by Ajmal and Koskinen (2008). One of the project managers described the way he established a common culture: "I wrote a 'working in the project' document with values and statements like 'avoid sending emails', 'call for short meetings when needed', 'talk to each other', etc.; then I worked with these values in the project to establish a culture of solving problems. Project communication was the major issue in the previous project in this area."

3.4.1 Competence Utilization

The utilization mechanism in the competence loop concerns how the organizational leaders utilize the competence of their employees in line with strategic goals. In this specific case, utilizing competence to attain organizational goals is carried out through either managing projects or other activities to develop or adapt IT solutions, or implementing IT solutions for different business units within the company, for which some internal but mainly external competence is used. Because of the critical financial situation, with several rounds of lay-offs, recruitment hardly exists. Instead, the external workforce is used to fill gaps or when additional competence is needed.

The process of allocating competence to projects is quite clear, because the project manager, together with the involved solution managers, breaks down the project and identifies which competences are needed. Projects are normally cross-functional, involving different functional areas and business units, and, in some cases, the customer. In a heterogeneous environment in terms of combining different competences, cultures, and business areas, the project manager has an important role. This specific case especially highlights the importance of matching the type of project with the weight of the different dimensions in the competence lemon to achieve an effective project management practice. Previous projects have struggled with matching the type of project with the project manager's competence because of a low understanding of the competence concepts.

3.4.2 Competence Accumulation

The accumulation mechanism in the competence loop concerns how the organization creates or acquires new competence. This could be through working on projects or other activities, or through absorptive capacity. The organization works in a cross-functional and multicultural environment. A business-oriented IT project normally involves resources and stakeholders from different business areas, markets, and countries. Barriers such as language, distance, time zones, and different competences impact people's ability to learn from each other.

Furthermore, the organization depends on specialists in different subject areas. The specialist could, for instance, be the solution manager for a business intelligence platform who works independently from the others who have other subject areas. In this example, the solution manager is responsible for managing suppliers and subcontractors, for which reason there is a low level of group learning. In addition, time is important for this organization, meaning that there is high pressure to deliver as fast as possible, which has an impact on quality and on time to reflect and share information with others.

Although the organization, to a large extent, relies on an external workforce, the most prevalent factor is how to absorb external competence. The manner in which the organization works to absorb external competence is mainly twofold:

- The first is documentation, in which the focus is on documenting how a solution has been implemented and why different solution decisions were made. The method of documenting has improved, but it still suffers from previous poorer documentation procedures. One manager described the situation: "Today we suffer from how badly we documented our solutions three to five years ago. With the improved documentation, we will have a better situation in two years' time."

• The second way is to work with handovers, which could be on-the-job training or different handover meetings with internal personnel.

A contradiction could be seen between projects and the parent organization, in which the parent organization works actively to absorb external knowledge, while the project focuses on project deliveries, or, as a project manager concisely expressed, it: "The project has no responsibility that the competences stay within the company." This comment could also be related to the high focus on time in projects.

3.4.3 Competence Assimilation

The assimilation mechanism in the competence loop is supported by individual solutions, wherein different functional managers follow up on performance in different ways. The high degree of external workforce utilization impacts how the functional managers interpret and understand new competences. They have a formal relationship with employees through performance appraisals and a more informal relationship with the external resources holding positions in the permanent line organization.

Because of the high degree of specialization among the resources, the functional manager not only has difficulties in interpreting and understanding the resources' technological knowledge, but also has a higher focus on the other dimensions of the competence concept. The specialists, on the other hand, value technological knowledge more highly and perceive that the functional managers have a low level of understanding of their competence. The latter gives a new perspective on what performance is and how to measure it. Even if a person develops deep knowledge within a specific area, other dimensions of the competence lemon are needed to apply the knowledge in a way that contributes to goals and objectives. Knowledge cannot be measured alone; the other dimensions need to be measured along with knowledge. Performance demonstrates how well knowledge and experience can be applied, and the other dimensions of the competence lemon control in which way knowledge and experience are applied in the specific context. Knowledge as such can be of value to the individual but has value for the organization only when it is applied in a way that contributes to the organizational goals and objectives. One aspect of this is the ability to share knowledge with others in the organization to support organizational learning.

In this organization, the projects normally have lessons-learned activities, sometimes several during the project. The purpose of these activities is to learn by looking at what occurred or what the project could have done differently. The project managers lead these activities, and they are carried out on a project level rather than on an individual level, with a view to following up on what the team members have learned during the project. A project team member put it as follows: "In the lessons learned, we catch up on what the team has learned but not what the individual has learned."

There are no other continuous or formal activities to learn from projects or interpret what individuals have learned during projects. The lessons-learned activities could benefit from adding learning to the agenda. One way of doing this is team reflection about what new knowledge has been acquired or if they have worked in a way that is more efficient than before. This kind of team reflection is also important for organizational learning, as they share knowledge in groups and learn from each other. The other aspect of lessons learned is that it is not an isolated activity at the end of the project; instead, it is performed continuously during the project. It is time well spent, which will add value to both the project and the organization.

3.4.4 Competence Transformation

In the fourth and last mechanism in the competence loop, transformation—which is described as a means of combining new and existing competence, reconfiguring core competences, and identifying competence gaps—the organization shows a weakness that may be due to a lack of understanding of the concept of core competences. The company has taken strategic decisions to outsource the development and maintenance of system solutions and move toward more standardized cloud-based solutions. This makes the identification of internal and external competence clearer and has led to a competence shift internally from system development and system architectural knowledge toward competences in managing suppliers, negotiation, and transforming business needs to requirements for standardized IT solutions. The high turnover of personnel and different waves of lay-offs have raised requirements for more effective processes and documentation to improve the transformation of knowledge from individuals to the organization, which Li (2012) and Oltra et al. (2013) describe as one part of organizational learning.

Another factor in the transformation mechanism in the competence loop that is important for organizational learning is promoting internal mobility. Internal mobility and job rotation are low in this organization for two main reasons:

- Firstly, the financial situation means that no one moves to another position through organizational changes, because the focus is on workforce reduction.

- Secondly, the specialist culture prevents people finding another job although they have expertise in a subject area and their expertise could be used elsewhere. A specialist focus has a negative impact on promoting internal mobility, even though it focuses on knowledge in the subject area.

The organization relies to a large extent on the external workforce in terms of contractors, consultants, and external suppliers. Several of the external resources hold positions in the line organization and act as employees in terms of responsibility and accountability. They are normally treated as employees except in terms of performance management and in the salary process. When the external resources leave, a great portion of their knowledge leaves with them, as explained by a project manager: "We have so many external resources. When the external resource leaves, the competence leaves." The high number of externals and the situation with several reductions in the workforce leads to the importance of a working transformation of the knowledge process.

3.4.5 Summary of Case 3

This case study showed that in an environment relying on an external workforce, the absorptive capacity and different ways of transforming knowledge and competence through processes and documentation become more important. Transformation of knowledge and competence has become even more important, because the number of employees has been reduced several times in the last couple of years. The onboarding process differs among different functional areas, but it is highlighted as important, although the turnover is quite highly dependent on the high proportion of external workforce utilization.

The organization has a high degree of specialization in working roles. Reduction of the workforce and the outsourcing strategy have led to the resources remaining in the line organization being specialized in their technological subject areas and alone in their roles. This situation impacts on knowledge sharing and on group learning, which is also evidenced in the negative correlation between specialization and learning. High levels of specialization impact negatively on learning owing to lower levels of group learning and knowledge sharing. This finding is in line with Cabello-Medina et al. (2011), who argue that knowledge, skills, and expertise tend to be depleted over time. The willingness to share will also decrease in a specialist culture (Starbuck, 1992), which can be seen in this case, in which the project managers have to focus on communication in the cross-functional projects to improve knowledge sharing in order to meet the project goals.

Another important factor in this kind of organization is its ability to understand what people have learned by working on projects or on other competence-creating activities. With a high level of external workforce utilization, the normal performance management processes do not work for two reasons.

- Firstly, the performance appraisal process, with goal setting once per year, is too slow compared to how quickly the surrounding environment changes. People are also working on several projects led by different project managers, so the functional manager has to capture individuals' performance in cooperation with project managers, who in turn are more interested in project delivery than in project team members' competence development.
- Secondly, the high proportion of external workforce utilization is not part of the performance management process, even if several of them have been working in the organization for many years. They are excluded from company-oriented HR activities such as performance management and competence development (Medina and Medina, 2014).

Moreover, the company operates in a financially stressful situation, with a reduction of workforce and requirements for system consolidation. The financial situation has led to the organization hardly working with recruitment or competence development in terms of external training. Instead, the focus is on delivering low-cost IT solutions for which time is important. The analysis shows that these factors have a negative impact on learning and innovation.

The high proportion of external workforce utilization in terms of contractors, consultants, and suppliers has several impacts on competence management.

- Firstly, it is difficult to maintain a common organizational culture with people coming from different companies bringing their culture with them. In this case, it is even more important that the project manager works to establish a culture in the project. One way of creating a project culture is to implement norms for communication within the project. Norms for communication can, for example, be how to meet, when to use emails, tools for communication, etc.

   Another way is to bring together project team members, or at least the core team, to clarify project tasks, collaboration approach, how to understand each other's competences, and similar areas.
- The second impact of having a high proportion of external workforce utilization is on how the organization can preserve knowledge when the external workforce leaves. Here, the project manager not only has a role to focus on project outcomes, but also has an organizational responsibility to keep knowledge within the organization. Methods can be documentation, hand-over activities, training sessions in the project, etc.
- The third major impact of external workforce utilization is on the performance management process, which normally only includes employees. This factor is closely related to preserving external knowledge within the

organization—the organization's absorptive capacity. Functional managers need to understand and interpret the external resources' competences in order to utilize their competence in the best way and preserve their knowledge when they leave.

This case also highlights the situation with high degrees of specialization. Specialists or experts have a tendency to do their job and leave. Project managers and functional managers need to include the specialists in a way that contributes to knowledge sharing and group learning. Both the specialist and the rest of the team will benefit from that. The specialist will benefit by understanding the whole picture, not only his or her subject area. The rest of the team, which obviously can include other specialists in other subject areas, will learn from the specialist and grow their knowledge in the area.

Other dimensions of competence will also improve when collaborating—for example, the ability to share and act in social interactions, to bring in other information to reach new conclusions, and many other dimensions of the competence lemon. In this case study, we could see that many people work alone, which also leads to a specialist situation. To prevent this, the organizational leaders need to establish a way of working in which people work together and share responsibility rather than being isolated in a working area on their own.

The strength in this organization was in the utilization mechanism in the competence loop. The project maturity was high, and project portfolio management was established and was working well. These two factors were the foundation for effective competence utilization, despite the financial situation.

This case is an example of an IT organization in a declining company, but the case also highlights the importance of context for the management of competence. Many of the findings can apply to other types of organizations, especially those which are highly dependent on an external workforce. In the next chapter, we will continue to discuss how different roles can benefit from the competence loop and the competence lemon.

3.5 Learning from the Cases

They develop their own language with their own terms and concepts,
and are using different vocabulary from us.

— Project team member

The first case was a case study of a public sector organization. The second was a case study of the R&D department in a fast-growing company. Finally, the last was a case study of an IT organization in a declining private company acting in a global market.

The three cases have several similarities as well as dissimilarities from a competence management perspective. What is significant is that the dissimilarities are not based on whether the organization is public or private, or whether it is a growing company or a company struggling with profitability. The differences among the cases is more related to context and organizational culture. We could see how the organizational culture impacted the application of competence and how different activities, such as group learning and sharing, were supported.

Another aspect is whether the organization is highly dependent on an external workforce or builds its business on employed co-workers. The following section will bring up the most common similarities and dissimilarities in order to explain how different factors impact on efficient competence management. It will be done from the perspective of the mechanisms in the competence loop. *[All references to the competence lemon can be found in Chapter 1; references to the competence loop can be found in Chapter 2.]*

3.5.1 Utilization

Competence allocation to projects is important in all three organizations, even if it works differently. In the public sector organization, there is an informal competence allocation process by which the project managers try to obtain resources previously known to the project and have to face the fact that the most competent resources have a high degree of responsibility for daily operation, meaning that the resources' commitment to the project may differ depending on the kind of daily issues that arise. In the R&D department, the product development activities are carried out in the functional organization, and the need for competence is known; in the IT organization, the project manager, together with solution managers, defines which competences are needed in the different projects.

Both private organizations work with some kind of strategic project portfolio, whereas the public sector organization does not have a formal project portfolio. The consequences of not working with a competence allocation process linked to a strategic project portfolio are expensive projects, poor quality, projects that are not completed, and repeated mistakes.

We can see that sourcing of external competence differs among the three cases. This factor is ranked low in the R&D department, because they have a low degree of external workforce utilization. The R&D department focuses on working with in-house competence, while the other two are highly dependent on external competence such as consultants, suppliers, contractors, and other external services.

The recruitment process differs among the cases. As a growing company, the R&D department has a well-developed recruitment process that also focuses on recruiting people in line with the organizational culture, while the IT organization hardly recruits anyone because the company is in the process of reducing the workforce. In the public sector organization, the respective functional manager is responsible for recruitment without any coordinated synchronization with other functions.

New resource introduction differs in meaning among the three organizations, even if the goal is the same—namely, to bring the new resource to perform effectively as quickly as possible. In the R&D department, the team has the responsibility to introduce new resources, and this is carried out mainly in two ways: mentorship and working with different tasks depending on one's previous knowledge level. The IT organization, on the other hand, has a high turnover of resources (both internal and external) and depends on high-quality onboarding documentation processes and solutions. Struggling with silos, the public sector organization's respective functional department introduces new employees to their assigned working tasks.

Because it is mature in project management methodology, the IT organization has well-defined project management competency models. The public sector organization is immature in project management methodology and has an unclear view of project manager competence. Finally, the R&D department has a strong functional matrix, in which the functional manager has more power than the project manager, and product development is performed in the functions, mainly using Scrum as a development method. These facts lead to weak project manager authority.

3.5.2 Accumulation

All three case studies highlight that trying new and challenging working tasks and working with problem solving enables a person to develop new competences. Furthermore, having the right attitude for work and being motivated are roughly equally weighted in the different case studies. Working in projects improves the efficiency of these factors, which is in line with Raj and Srivastava (2013), who show that people develop their problem-solving abilities and accomplish tasks faster when they are working in teams. Thus, we can conclude that a higher level of collaboration will improve a person's problem-solving capability as well as their motivation and attitude to work. Workplaces should be designed to support collaboration to create an innovative and knowledge-creating environment.

The difference among the case studies is mainly in three aspects:

- Firstly, to what extent the organization works with external resources
- Secondly, the amount of cross-functional work or number of projects
- Thirdly, whether the organization is team- or specialist-oriented

The correlation between the degree of working with external resources and its impact on generating new competence is clear when comparing the three cases. In an organization that is to a large extent dependent on external resources, such as the IT organization, the ability to absorb external knowledge is valued more highly than in an organization in which all resources are mainly employees, such as in the R&D department. The public sector organization is between the other two, which is also reflected in the importance of absorbing external competence. An organization that is highly dependent on an external workforce needs, as Wang and Ahmed (2007) put it, a strong ability to learn from external parties, integrate absorbed knowledge, and transform it into organization-embedded knowledge.

The IT organization also acts in an environment with many business units, with teams located in different parts of the world and with different external suppliers. This makes cross-functional collaboration a factor that is considered as a developing competence with higher importance than in the case of the other two organizations. Project teams with members from different functional parts of the organization are likely to achieve more effective results and solutions to complex problems than teams from a single functional area, comparable to the empirical evidence from Rynes, Colbert, and Brown (2002), who found that cross-functional teams with members from different areas show higher performance and positive outcomes in terms of project and product quality. Cross-functional collaboration is similar to a heterogeneous environment, which is a factor that has an impact on generating new competence, as discussed in Chapter 1. In summary, a heterogeneous team is more productive, but the leader has to establish a team culture supporting collaboration and togetherness, in which the team members respect each other's competence and work toward common goals, sharing knowledge.

The third major difference is whether the organization has a team- or specialist-oriented culture. In the case of the IT organization, the specialist culture is obvious from three perspectives:

- The first perspective is the traditional expert role, in which the specialist has deeper knowledge within a specific subject area. In this case, the specialist has a tendency to fly in, do the job, and leave. The risk in this case is that the specialist does not have the full picture of the goal of the project and does not do the right thing. They neither share their knowledge with the other team members, nor learn from them. This specialist perspective is related to heterogeneous teams, as discussed above.
- The second perspective of the specialist role is when a single person has responsibility for a subject area. In this case, the specialist does not need to have expert knowledge in the area, but there is no one else to interchange

ideas or discuss issues with. The subject matter area responsible will be isolated, with a low level of knowledge sharing and group learning. The risk in this case is that the specialist's knowledge will eventually be obsolete or that he or she will take decisions not in line with the rest of the organization. It would be more efficient to broaden the subject area and share responsibility among a greater number of people.

- In the case of the team-oriented organization, team loyalty has a tendency to be strong. This can have both positive and negative consequences:
  o The positive consequences are that the team members know each other, know each other's competences, and can in this way take on new tasks and divide them among the team members in an efficient way.
  o The negative consequences are that, in many cases, team members do not use the competence in the organization in the optimal way. They might break up teams, because they have new projects that require teams with different experiences and competences. A team with few new influences has a tendency to be less efficient over time.

3.5.3 Assimilation

The most important factors in the assimilation mechanism in the competence loop in all three case organizations were performance management and interpreting new competences. Both factors are a way for managers to measure, assess, and interpret individuals' competence. Performance management is the formal process, normally carried out through a performance appraisal process, whereas interpreting new competence is an informal process through which managers try to understand and interpret individuals' competence on a frequent basis.

Lopez-Cabrales et al. (2010) argue that an organization with a clear development program will result in employees tending to match their competence to the needs of the organization. This was supported in all three cases studies, which showed an absence of effective competence development programs and difficulties linking individual performance to the overall business goals. This led to a situation in which the employees had difficulties seeing the value of the performance management process. The formal process, with goal setting once a year and follow-up after six months, was also considered inefficient, because people were taking on new tasks during this period and goals became out of date. The traditional performance appraisal process is out of date for knowledge workers.

Interpreting new competence refers to an informal process that interprets individuals' competence on a frequent basis. An individual's competence is interpreted and assessed in different ways in the different organizations, as well as within an organization. Feedback on individuals' performance will lead to better competence development, which is supported by Kim and Lee (2012) and

Lopez-Cabrales et al. (2010), who argue that continued feedback on individuals' performance will have a positive effect on the value and uniqueness of human capital. With continuous performance measurement and follow-up on people's competence, the organization will become more efficient, which will benefit both co-workers and organizational leaders. Modern knowledge-intensive organizations should apply an agile performance management, in which performance and competence are followed up on a frequent basis.

Three factors are related to projects—namely, project manager feedback, poor performance, and learning from projects.

- Employees in project-intensive organizations spend considerably more time with their project managers than with line managers, which leads to line managers being at a distance from the people that they are responsible for leading. The project manager should have a role in HRM practices, such as performance management systems in project-intensive organizations, but today that role is not clear. In the studied organizations, feedback from the project manager is a factor in the assessment and interpretation of project team members' competence.
- One issue raised was the tendency to feed back on poor project team members' performance more often than on acceptable, good, or great performance. This can probably be explained by the project reporting focusing on risks and problems rather than people's personal development. Based on the above reasoning, performance—good or bad—should be a part of project manager feedback.
- Learning from projects is another factor in understanding and interpreting what competences the project team members have developed during the project. In general, activities such as lessons learned and retrospective meetings focus on the project management process rather than on what project team members have learned during the project or within the last Scrum sprint. To be more efficient in managing competence, learning from projects should include the competences project team members have acquired during the project, and these learning activities should be performed on a frequent basis, not just at the end of the project, when both the project manager and the project team members are most often working on other projects or other activities.

3.5.4 Transformation

All three studied organizations struggled with identifying their core competences, which need to be identified and defined in order to be competitive in the market. Core competences are those capabilities that are critical to a business's achieving a competitive advantage. The strategic future of an organization depends on the organization's ability to utilize its competences in relation to the organizational goals—independent of whether the organization is private or public and whether it is at a company, division, or department level (Eden and Ackermann, 2010).

Furthermore, core competences can be identified by analyzing the relationship between competences and the organization's purpose and business goals. One way to find the link between an organization's business goals and its competences—which defines the organizational core competences—is to describe the organization's potential successes and failures and, in this way, analyze the competences needed to support the success. In a knowledge-intensive context, learning should be considered as a core competence, in that this allows continuous generation of new knowledge.

The above reasoning affects the factor of identifying internal/external competences, because it is the organization's identification of core competences that decides which competence to source from suppliers and which to develop within the organization. Here we could see that the different organizations studied showed a wide spread in terms of strategy for utilizing external competence, but the decisions on what to source externally were based more on the context in which they were acting than on strategic decisions from a core competence perspective.

How to transfer new competence to the organization—an important part of organizational learning—differs in the studied organizations. In general, this is mainly based on how the organizations view process and documentation. Tacit knowledge is ambiguous in nature and difficult to duplicate, whereas explicit knowledge is easy to store and communicate and will not be lost because of employee turnover.

3.5.5 Competence Transfer in the IT Organization

The IT organization is the most mature in documentation and process development and has a high turnover. In this case, projects work according to traditional project management methodologies, in which formal documentation is required to a larger extent than in projects working according to agile project management. People put high levels of effort into documentation and process development, through which they convert individual knowledge into organizational knowledge, in what Nonaka (1994) calls *externalization* and Jasimuddin (2014) calls *codification strategy*.

3.5.6 Competence Transfer in the R&D Organization

On the other hand, the R&D organization has a low turnover of people, is recruiting many new employees, and is working according to agile project management.

This organization puts less effort into documentation and process development, instead focusing on group learning and sharing of knowledge, in what Nonaka (1994) calls *socialization* and Jasimuddin (2014) calls *personalization strategy*. If the turnover rate in the R&D organization increases, the risk that the knowledge will be lost increases. The minimal documentation in agile project methodologies, in combination with the culture, makes documentation a low priority area; as a project manager put it, "We are not good at documentation: it is a part of the culture."

Promoting internal mobility is regarded as an important factor for the transfer of competence within an organization. The R&D organization works in a functional matrix, in which the functional departments are strong. The powerful functional departments lead to few incentives for career development and a situation in which few people move between the different functional departments. A low level of movement between the functional departments leads to less transfer of knowledge within the organization, which is in line with Bredin (2008), who sees a need to have career paths for functional specialists in project-intensive organizations to retain skilled and valuable specialists, develop necessary competences over time, and promote people moving across organizational boundaries.

Figure 3.1 describes the three organizations from two perspectives: the knowledge intensity (X) and maturity in competence management (Y). On the maturity axis, three levels are specified: (1) *establish*—meaning that there is a lack of several competence management factors; (2) *develop*—meaning that most of the competence management factors are in place, but several of them need to be developed; and (3) *optimize*—meaning that almost all of the competence factors are in place, but there are opportunities to optimize them.

The method of measuring organizations based on their knowledge intensity and maturity in competence management gives the baseline for deciding on

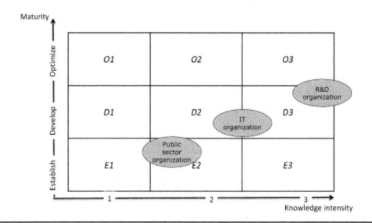

Figure 3.1   Knowledge Intensity–Maturity Graph

how to approach improvements to establish a way of managing competences in line with the organizational goals and strategies.

In Figure 3.1, we see that the R&D organization is the one with the highest knowledge intensity, which is natural because it is working with research and development. It is in the upper right corner at D3 (development maturity level in a high knowledge-intensive organization), and is the most mature organization from a competence management perspective, mostly as a result of three things:

1. The organizational culture is strong and functional.
2. The management team promotes and encourages knowledge sharing and innovation.
3. It has a clear strategy and roadmap process that facilitates transparency as to which projects will start and which will not.

The IT organization is lower on the knowledge–intensity axis than the R&D organization because it works with a degree of repetitive work. It is placed between D2 (development maturity level in a medium knowledge-intensive organization) and D3, but is lower in maturity than the R&D organization. Its strengths are in its project maturity and project portfolio management, but it suffers from mechanisms to generate new competence and strengthen competence owing to its specialist culture, low degree of sharing, high level of external workforce utilization, and non-functional organizational culture.

3.5.7 Competence Transfer in the Public Sector Organization

The public sector organization has both the lowest knowledge intensity and the lowest maturity in terms of competence management. It falls into E2 (established maturity level in a medium knowledge-intensive organization), meaning that it lacks several factors to be effective in competence management. Its weakest mechanism in the competence loop is utilization, and, because of that, it lacks project portfolio management, has a poor competence allocation process, and struggles with project governance. In addition, its project maturity is low, and the organizational culture prevents effective competence management.

3.6 Conclusion

In summary, we found both similarities and dissimilarities in managing competence when comparing the different organizations. We can conclude that all the different factors constituting the mechanisms in the competence loop exist in all kinds of organizations, but they are of various weights depending on the focus of the organization. Thus, we can conclude that, in different contexts, we need to put higher effort into different factors in the various mechanisms.

In the next chapter, a methodology for generation of new competence—called REPI (Reflection, Elaboration, Participation, and Investigation)—will be introduced. REPI is linked to the competence lemon and can be used in the accumulation and assimilation mechanisms of the competence loop.

Chapter 4

REPI

Figure 4.1   REPI: Reflection, Elaboration, Participation and/or Practice, Investigation

This chapter will describe a methodology for generation of a new competence called *REPI* (see Figure 4.1). In the chapter, the six dimensions of competence from the competence lemon will be linked to different activities within the REPI methodology. All references to the competence lemon can be found in Chapter 1, and we recommend reading Chapter 1 before reading this chapter. We also reference the competence loop in this chapter, which is described in detail in Chapter 2.

4.1 The Basics of REPI

Over more than two decades, Dr. Alicia Medina has developed a methodology called REPI, a learning methodology that can be used to generate competence, not just to acquire knowledge. The reason for developing this methodology was her insights based on field experience that were in line with the learning paradigm introduced by Barr and Tagg (1995). This was a new way of seeing the mission of higher education and, by extension, education in general, instead of the traditional way of instructing or teaching to produce learning. Learning, according to REPI, is more than a transfer of knowledge from the teacher or trainer to the students; it is about active participation, during which knowledge is constructed, and the students take ownership of the newly gained knowledge.

During Dr. Alicia Medina's work as a manager in different organizations and corporations among other engineers, she tried to understand why some engineers recalled very little of certain areas that they had studied for years. At that time, she found the Cone of Learning that was created by Dale in 1969. The Cone of Learning was based on his previous work on the Cone of Experience from 1946. The cone represents the ability to recall information that has been previously taught. According to this research, people recall 10 percent of what they read, 20 percent of what they hear, 30 percent of what they see, 50 percent of what they see and hear, 70 percent of what they say, and 90 percent of what they say and do. Dale also distinguished between passive and active learning: reading, hearing, and seeing are passive learning, while saying and doing refer to active learning. Active learning can be compared to learning by doing, which is one of the components in the accumulation mechanism in the competence loop.

Inspite of Dale's research being old and having been criticized in more recent studies—because the percentages can have other explanations, such as motivation, interest in the topic, or other context-related influences—the principles are still highly valid. The essence of those results is giving importance to active participation and minimizing or eliminating passive methods when it comes to creating competence.

Bringing in Dale's Cone of Learning to the modern competence loop, we can establish the new learning curve model presented in Figure 4.2. The learning curve visualizes the different steps of learning going from passive to active learning which generates new competence by the concept of learning-by-doing.

In addition, Bloom's taxonomy from 1956 was also studied. This taxonomy consists of six elements: *knowledge, comprehension, application, analysis, synthesis,* and *evaluation.* In 2002, Krathwohl published a revision of Bloom's taxonomy, developing the following categories: *remember, understand, apply, analyze, evaluate,* and *create.* Other authors, such as Male, Bush, and Chapman (2011), Hung, Choi, and Chan (2003), and Passow (2012), further developed this view and called for a change in the educational system. That is what REPI intends to do.

REPI is composed of four modes that need to take place in order to guarantee the learning process. Even though REPI was originally developed for teaching purposes, during the years Dr. Medina and her colleagues from the consultancy firm Quini Consultant in Sweden, including myself, started to use it for coaching and competence development and as a technique for performance management and team building.

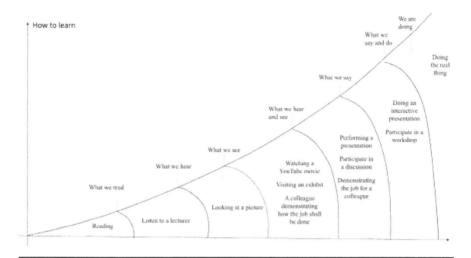

Figure 4.2 The learning curve adapted from from E. Dale, *Audiovisual Methods in Teaching*, 1969, NY: Dryden Press.

The modes constituting REPI are *reflection, elaboration, participation and/or practice,* and *investigation.* Those four modes, with corresponding activities, can be used in different ways, depending on the purpose for which REPI is being used. Performing a mapping of REPI and the competence lemon presented in Chapter 1 showed that there is congruence between them. Furthermore, REPI supports the competence loop framework presented in Chapter 2. For these reasons, REPI will be referred to in the different chapters in this book. The following sections will describe each of the modes in more detail, followed by some practical uses for the methodology.

4.1.1 Reflection

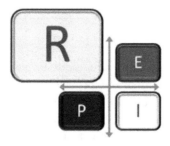

Figure 4.3 REPI: Reflection

Reflection (see Figure 4.3) is the mode that refers to assimilating and constructing one's own thoughts and opinions about a given topic or situation. In this mode, associations with previous knowledge and experiences are created, and the individual generates an opinion about the topic. As described in Chapter 1, reflection is one of the most important factors in generating new competence. People can reflect on mistakes, successes, new insights, what other people have said, their own presentation, or anything else. In the reflection mode, previous activities are considered and looked at from different perspectives.

Furthermore, in the reflection mode, people work with the lemon wedges that refer to the following dimensions of the competence lemon: personal capability, social capability, ability to manage complexity, and ability to learn. By reflecting on their own behavior, people gain new insights into, for example, how to be more pedagogic, innovative, better at communicating their thoughts, etc. The ability to learn will improve through reflection on how to take in new knowledge. Finally, when a person reflects, for example, on how people interacted with one another, how decisions were taken, how they acted in different situations, or how different topics relate to each other, then the ability to manage complexity develops.

4.1.2 Elaboration

Figure 4.4   REPI: Elaboration

Elaboration (see Figure 4.4) is the mode or activities in which, with some given information or a number of facts, the mind elaborates, making assumptions, interpretations, and associations with other topics. In addition, elaboration can be defined as being able to take one idea and embellish it. The focus in elaboration is on adding details to create a logical and comprehensible order among the information that is given. This mental process is highly selective and is congruent with people's own preferences, current needs, and/or previous knowledge. Moreover, it is about letting the mind fill the gaps in the information provided with no other external search. In the elaboration mode, the following dimensions of the competence lemon are present: personal capability through problem solving, ability to learn, and ability to manage complexity.

4.1.3 Participation and/or Practice

Figure 4.5   REPI: Participation/Practice

Depending on the purpose and context in which REPI is used, the *P* stands for either participation or practice, or both (see Figure 4.5). Participation is the mode in which people share knowledge or interact with others in a team, meaning that they are actively participating in a discussion. Participating is about knowledge sharing and interacting with others, expressing and discussing opinions. Practice, on the other hand, is about testing ideas; it is about doing and putting into practice the newly gained skills or insights.

When using participation, group learning and knowledge sharing are being utilized to generate new competence. While practicing, new competence is being generated by using problem solving or trying new and unknown things.

In the participation or practice mode, all the dimensions of the competence lemon are present. New knowledge is gained by practicing ideas or thoughts. By actively participating in discussion, we improve both our social capability and our personal capability by presenting and sharing our thoughts in different ways.

Another view of this mode is that new knowledge and experiences are linked to our cognitive mental map and, in this way, impact on our ability to learn and to manage complexity is improved. Finally, by sharing ideas and practicing new leadership styles, we improve our leadership qualities.

4.1.4 Investigation

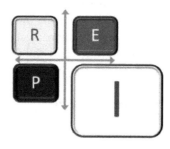

Figure 4.6   REPI: Investigation

The investigation mode (see Figure 4.6) is about searching for information, facts, and experiences beyond the current state and situation. Moreover, it is about seeking to use different sources, such as the internet, literature, benchmarking, competitor analysis, best practices, cases, and standards. In addition, investigation is about conferring with stakeholders or others with experience within the given subject matter area. In the investigation mode, a person works with all the dimensions of the competence lemon. New knowledge is gained by seeking information from different sources. This information is then linked to and/or integrated with previous knowledge and, in this way, new insights are generated. This is what the dimension from the competence lemon termed *ability to manage complexity* is about. Finally, the *ability to learn* dimension is also improved by gathering information and linking it together.

4.1.5 Processes Based on the Modes

The four combined phases of REPI can also be seen as different processes for which:

- Reflection and participation (RP) is the internal mental process in which people reflect on insights and what is happening in groups.
- Elaboration and investigation (EI) is the process of filling the gaps and finding new insights. People seek answers to questions in order to fill the gaps. Information is gathered from different sources and elaborated on to find answers.
- The combination of practicing and investigation (PI) is the process of external interaction, during which people practice new insights and ideas and, together with others, seek more information.

- Finally, the use of practicing and reflection (PR) is the process of making a stance and linking new insights and experiences. In this process, people practice new ideas and proposals and reflect on the result, leading to conclusions and standpoints.

4.2 How to Use REPI

REPI can be used in various ways and for different purposes. Because there are no given sequences of the various modes of REPI, the modes can be used differently in different situations. This section will present how REPI can be used depending on purpose and context.

4.2.1 REPI Meeting: A Case Study

A REPI meeting is one in which the REPI methodology is used and a series of techniques takes place in order to increase the group's competence on a specific topic. It usually has a goal or a pre-defined expected outcome and is managed by a REPI facilitator. The following case from real life illustrates how a REPI meeting could be used.

A project team in a large corporation is working on the task of finding a new recruitment system that will replace the existing one. People from different departments, with knowledge and experience from various subject areas, are allocated to the project. They are gathered in the very first meeting.

In this meeting, the project manager has the role of REPI facilitator. She starts the meeting with the reflection mode by asking the team, "What do we want to achieve by selecting a new solution, and what are the pains and the well-functioning things in the current solution?" Furthermore, she presents some facts about the number of recruitments that are made currently, the time it take to close a recruitment case, and the average cost associated with recruitment. Based on this information, the team is given some minutes to reflect on the questions and those facts.

In the next step, the REPI facilitator asks the team to share their reflections and asks all the participants to share their thoughts and opinions about the reflections being presented. This is the participation mode from the REPI model. An important thing at this stage is to facilitate the conversation and to move the conversation by having a constructive, open, and genuine interest in the topic.

Then next step is the elaboration mode, which is introduced by giving the participants some minutes to work in smaller groups or individually, using a technique called *parallel thinking* (de Bono, 1994). This technique means that the groups work in parallel with the same question or task, but first the information provided needs to be further developed by each group, meaning elaborating on the theme from the starting point that each group chooses to focus on. In this case, the team worked with three main questions: (1) Where do we want to be when the project is completed? (2) What are the business needs that the new solution will fulfill? (3) What are the more important aspects to be taken into consideration?

One of the groups elaborated on the financial aspects and the improvement in respect of time to recruit. Another group chose to elaborate on the usability of the solution, stating a series of requirements that would secure happy users and a requirement for less support. Finally, the third group focused only on the collaboration aspects that were needed for the new solution, such as notifications between the recruiting manager and the recruiter, the visibility of the status of the recruitments, etc.

When the elaboration period was finished, a new participation session was initiated, and all the groups shared their thoughts. This session led to a picture of the future that covered several aspects that were elaborated on depending on the preferences, knowledge, previous experience, and, of course, the roles of the participants. In this case, the elaboration session was successful and helped the team to cover and understand several aspects of the same task. However, there were cases in which some team members felt insecure and wanted, or even demanded, that the REPI facilitator (who was also the project manager) give them precise or more concrete instructions. This is one of the challenges that the facilitator needs to face and be prepared for—being able to explain that there is a purpose in giving few instructions, because giving detailed or precise instructions will limit the mental processes of elaborating, being innovative, thinking outside the box, and having different groups or individuals covering different aspects.

At the end of the meeting, the REPI facilitator gave the team members the task of searching for solutions, reading the industry reports on recruiting, contacting some consultancy firms and asking them to share their experiences, and benchmarking what other companies have done in the same area. This meant entering into the investigating mode.

The next REPI meeting started with a participation session, during which the team members shared the results of their investigations. This was followed by a reflection period on both how the investigation had been performed and the results. After that, the REPI facilitator presented a new task on which to elaborate.

A REPI meeting usually takes around one hour, and the last couple of minutes are always dedicated to sharing the learning acquired during the meeting.

4.2.2 Teaching

As stated above, REPI was developed to fulfill a teaching purpose, as a methodology to learn in a way that was different from the traditional way in which a teacher or trainer is simply transferring his/her knowledge to the students.

During lectures, REPI is usually practiced by having the teacher present some facts, theories, or questions, then giving the students some minutes to think and elaborate on the topic in a spontaneous way, without seeking more information from sources other than their own minds and their own thoughts (elaboration). This can be done individually or in small groups, depending on both the number of students and the shape of the classroom. On some occasions, the group may be divided, and they can be asked to create a presentation of their results.

Then the students, either individually or with one representative from each group, have a short moment to share their results with the others, and the others can ask questions and share their opinions on the reflections being presented (participation).

As a next step, the students can be asked to analyze the different presentations and to reflect on them, sometimes in quite an open way, reflecting on the presentation as a whole; in other cases focusing on a specific area—for example, cost, risks, opportunities, etc. (reflection). As a next step, the insights based on the previous reflection period can be either presented again for the whole group, or assigned as homework that needs to be handed in to the teacher or to an opponent group. A further next step could be to continue searching for more examples, evidences, or other real-life examples and/or experiences (investigation).

There is not a fixed order for REPI. In some cases, the starting point can be the investigation mode, by searching and then sharing (participation) followed by time to reflect (reflection). Otherwise, the elaboration mode with some facts is the starting point, followed by searching for more information (investigation). After searching for more information, people reflect on the results (reflection), after which the final results are presented and discussed within the group (participation).

4.2.3 Coaching

REPI is a powerful method for coaching. A coaching session can be considered as a REPI meeting as explained above, but having only two participants—the coach and the person being coached.

The coaching session can start by reflecting on a particular topic and by sharing reflections and questions. The coaching session can be followed by elaboration on a possible solution that the person needs to put into practice. The next coaching session could start by sharing the experience of what has been put into practice, followed by reflection on how the new experience could be improved or be incorporated as part of normal procedures.

Depending on the purpose of the coaching, the person being coached could be asked to read a book that will be discussed later or to visit another company in order to gain new knowledge (investigation). There are many ways of managing the coaching, but the basics are that the people should reflect, elaborate, practice, and search for new information and knowledge.

4.2.4 Competence Development

Using REPI for competence development by an individual or a group follows the same path as presented in the previous sessions—the REPI meeting or REPI coaching session takes place and the goal of the session is to gain competence about a specific topic, area, or situation. In order to acquire lasting competence, it is important that practicing takes place, because doing is key to acquiring competence and not just knowledge.

In the real-life example of the REPI meeting previously presented above, the goal was to gain competence about both the reason for implementing a new recruitment solution and recruitment solutions in general. This was an example of people gaining new knowledge and competence in a group. How REPI can be used for individuals is further exemplified in Chapter 6.

4.2.5 Performance Management

REPI is a natural tool for performance management if the manager acts as REPI leader for the employee. This will establish a natural way for the line manager and the employee to discuss competence and performance without creating tension. It will also be a continuous performance management approach, which is exemplified in Chapter 6 and called agile performance management.

4.2.6 Team Building

When using REPI as a technique for team building, the focus is on the participation mode, because participation is about sharing and giving an individual, at the time, the space to "own" some minutes to express his or her thoughts, experiences, or arguments. It also allows the rest of the team the possibility to discuss whatever was shared. This is a very powerful way of creating an understanding of each other's preferences, and it gives everyone an opportunity to express themselves and the freedom to choose what and how much to share. The same applies when reflection is followed by participation: the team members become aware of both similarities and dissimilarities in each other's ways of analyzing a problem or reflecting on a topic. When reflection is performed in small groups, it also helps to train the ability to work in a group and to follow others' thoughts. Working with the elaboration mode in a group, in which participants have to sort out different ideas and to agree on some aspects to achieve a consensus, creates togetherness in a team.

Investigation is also used when the team needs to search for more information together—for example, by interviewing an external expert, or visiting a competitor. In this case, investigation, in the same way as practicing, is an opportunity to perform a joint activity, developing collaboration skills.

4.3 Final Thoughts

But perhaps the power of REPI lies in the fact that it gives the team members a technique to learn, to know, to understand, and to gain insights about each other. A team that uses REPI meetings on a continuous basis utilizes each meeting as a learning arena, and as an arena for sharing, for innovative thinking, and for collaboration.

In this chapter, a methodology for generation of new competence was introduced. This new methodology, together with the competence lemon and the competence loop, forms a foundation for effectively managing competences. In the next chapters, we will take the perspective of project management and the role of the PMO symbolized by how to fly a kite, followed by how different roles in the organization can benefit from the competence lemon, the competence loop, and REPI.

Chapter 5

The Project Management Kite and the PMO Role

By Dr. Alicia Medina

I am honored to be invited to write some of my thoughts and metaphors about project management and the roles of the PMO. It is my hope that this way of illustrating and comparing project management with a kite will contribute to creating a better understanding of the importance of considering different aspects of project management and will create discussions in which frameworks, methods, and tools are present but decrease in dominance in the field. Furthermore, for any organization to become projectified and to be successful, both the sponsorship and the steering group need to allow the PMO to create project management and projects to point it in the right direction.

Dr. Alicia Medina
Malmö, Sweden
July, 2017

5.1 The Role of the PMO

In a projectified or project-intensive organization, there are different aspects of project management that a project management office (PMO) needs to manage. These need to be managed even in organizations that do not have a PMO:

someone needs to be responsible for these aspects or the organization will not gain the benefits expected from being projectified.

These aspects are classified as follows:

1. Framework, methods, and tools
2. Governance
3. Leadership
4. Domain and context

- The first classification is about the *frameworks* such as the Project Management Institute (PMI) and managing successful programmes (MSP), program and project management for enterprise innovation (P2M); and *methods* such as praktisk projektstyrning (PPS), projektet för projektstyrning (PROPS), MiniRisk, Lichtenberg, Monte Carlo, economical value added (EVA), and earned value management (EVM). It can also be a mix of those frameworks and methods, such as PRINCE2®[1] (Projects IN Controlled Environments). Among the *tools* are MS-project, Projecplace.com, time reporting tools, etc.
- The second classification covers *governance,* meaning how the projects and programs are controlled and steered. In this classification, we have aspects such as strategy, forums, steering groups, councils, policies, processes, stakeholders, and control mechanisms.
- *Leadership* aspects such as motivation, conflict handling, negotiation, cross-cultural diversity, and decision making constitute the third classification of project management that a PMO needs to manage.
- The fourth classification is composed of both the *domain* and the *context* in which the project will operate. The domain is, for example, IT, retail, pharma, construction, or finance; but within domain we also include project tasks such as roll out, outsourcing, business transformation, new solutions, and decommissioning. The context is about aspects surrounding the project, but also aspects that are internal to the project. Here we make a distinction between *context* and *environment.* The context consists of aspects such as downsizing, growth, in-house, offshore, global, multicultural, and new business.

5.1.1 Why Consider Project Management Aspects as Part of a Kite?

In order to understand the use of a kite as a metaphor for project management, we need first to present the components of a kite and its function.

---

[1] PRINCE2® is a registered trademark of AXELOS Limited. All rights reserved.

The cover of the kite is called the *sail*. The two small areas at the top of the kite are called *pilot* or *leading sails,* while the two larger segments are known as *driving sails.* The pilot sails partly control the direction in which the kite moves, and the driving sails provide most of the lift. In the project management kite, we place leadership and governance on the pilot sails and framework, methods, and tools, along with domain and context, on the driving sails.

The four aspects need to be present and need to be taken into consideration, in addition to their position/role. PMOs that choose to put the framework, methods, and tools at the front and try to use them to steer the direction of project management will not be successful.

Framework, methods, and tools are aspects that are enablers but can never control the direction.

If the domain and context are ignored or only partially taken into consideration, there is a risk that project management is focused only on *administering* a project and not *managing* a project. This is present in many less successful PMOs, in which there is a belief in having generic project management and a practice of "one size fits all."

A kite can be considered as a sort of aircraft that is held by a string. The string is used to stop the kite from flying away with the wind. If we let project management "fly" in the organization without taking responsibility for the direction, it might fly away with no possibility of being directed or redirected. This is one of the main tasks for a PMO—to be responsible for the direction of project management, for the continuous improvement of practices, and for competence development in the area.

Furthermore, because a kite has no engine, it needs something else to make it move through the air. Just as the power source for a kite is the wind, the power source for project management is the organizational climate, the readiness, and the commitment from upper management. In days without wind moving over the kite it won't fly: this is what happens with project management if the climate in the organization is not appropriate for running successful projects. Project management will remain something that will not be adopted, or will be only partially adopted.

Everyone who has worked with kites will agree that some kites need lots of wind, while others need very little wind for them to fly; the explanation lies with the construction and having the right proportions. The proficiency of the one holding the string is also a determining factor. The same applies to the project management kite: if it does not have the right construction—meaning the right proportions between the four aspects of project management that a PMO needs to manage—or if the responsible PMO does not have the necessary skills and competence, then no matter how the organizational wind is blowing, it will be difficult to accept project management.

5.1.2 What Makes a Kite Fly?

Just like a normal airplane, or even a bird, there are four forces that affect a kite when it is flying. These are *gravity, lift, thrust,* and *drag.* A kite is affected by gravity, a force that pulls everything down. In an organization, this means that if the people who compose the organization have a tendency to have the same power as gravity when it comes to change, projects will probably fail.

If project management is introduced in a way that is perceived as something difficult and heavy, people in the organization will pull it down. That is why it is important to keep in mind that the heavier a kite is, the harder it will be to fly. The challenge is to understand what the proper weight is. This is what makes some PMOs successful while others die or are lethargic. Heavy implemented frameworks and methods, or heavy governance aspects, will affect the fly. I have seen organizations that start a PMO and put in a lot of effort in both monetary terms and time, but the organization feels the "heaviness" of what it is delivering and does not accept the delivery, leading to a loss of legitimacy for the PMO. In addition, I have seen the effects of this heaviness, where initially successful organizations have moved into an unsuccessful stage: the negative change occurs when they move from a functional to a projectified structure. In this change, the PMO has gained power and put pressure on the organization instead of being supportive, which has led to the organization's going back to performing line activities instead of projects.

Thrust is a force that makes things move forward through the air. In the case of a kite, it cannot produce its own thrust: it needs to be held in place by the string while the wind moves past it, thus generating the thrust. That is why the PMO is responsible for holding the string. But there is another similarity. When there is no wind, a kite will only fly if the person holding the kite string starts running, generating his or her own wind, as symbolized in Figure 5.1. This is what a PMO needs to do in many cases if the organization is not ready to adopt projectification as a way of working. In such cases, the PMO needs to create readiness and acceptance. Moving things forward in terms of project management can be done through different change management activities and, of course, by training people. It can also be done by coaching, mentoring, supporting, and showing that there is a winning concept.

When there is wind, the PMO can somehow "relax" and just hold the string: project management largely flies on its own but is given a supporting hand when needed, as visualized in Figure 5.1.

Is it then possible to say that having the proper weight and construction is enough to get the kite to fly?

The answer is no, because kites are affected by a kind of friction called drag. This friction is what helps the kite to remain in the sky and not just fall to the

Figure 5.1   The forces that make a kite fly.

ground. Even if the kite itself generates some drag, there is a need for extra drag, and this is provided by the tail. It is also the tail that allows the one holding the kite to point the kite in the right direction.

So what is the tail in the project management kite? Is the tail of project management constituted by the organizational sponsors, or it is the members of the steering group?

To answer that question, we of course are going back to the kite. In the same way that a kite tail has more effect and generates more drag if is constituted of several lengths instead of just one long piece, the project management kite will be more effective by having a series of decision makers or steering group members, which will also create stability in the project.

Does this mean that, in the case of many sponsors or steering group members, the kite will fly better? The answer is no, because if the kite has too much tail it will be stable, which is positive, but it will probably be difficult to keep it flying because of the extra weight caused by the excess tail. That is why the tail is also a challenge from the perspective of project management. A project normally has many stakeholders and one sponsor. However, in many cases a project has a few or several stakeholders with high power and a positive high interest in the project: in this book we will call them *HiPi* (high interest, high power interest). HiPis are in most cases decision makers that want the project to be successful and will stand behind the project manager and support him or her in making decisions regarding the project. HiPis can potentially move from being positive to being negative toward the project, or the other way around. The conclusion is that HiPis tend to be interchangeable, in the sense that they can support a project but suddenly switch goals.

Does this mean that in the case of many HiPis or steering group members, the kite will fly better? The answer is no, as described above—if the kite has too much tail it will be stable, which is positive, but it will probably be difficult to keep it flying because of the extra weight caused by the excess tail. That is why the tail is also a challenge from a perspective of project management. Is it enough to have three, six, or ten HiPis in a project? The answer will depend on the type of organization and project and on the history of the organization, but it will never be appropriate to have more than six HiPis: the organization will be too heavy and the project will have a problem flying. The parts constituting a PMO kite are summarized in Figure 5.2.

Figure 5.2　The PMO kite.

It is not recommended to fly a kite in a storm or while it is raining, because it will not fly successfully. The same applies to project management in general, and to particular projects where all conditions are against the project. If the conditions for managing projects in a proper way not are possible in the organization, it does not matter how well the four aspects of project management are designed and managed: it will not fly successfully.

I wish you happy flying with successful projects!

5.2 Conclusion

Having taking the perspective of a PMO, in the next chapter we will look into how different roles in the organization can benefit from the competence lemon, the competence loop, and REPI.

Chapter 6

Competence Management in Practice

A project is a gap with risks and opportunities.
You learn from these.

— Project team member

In the previous chapter, we looked at project management from the project management office (PMO) perspective, presented and symbolized as a kite. The PMO can have different shapes and responsibilities, depending on the organization and how it is implemented. In this chapter, we will look into different practical perspectives of competence management and how the competence lemon, the competence loop, and REPI* can be used in different situations. Because the context is knowledge-intensive, project-intensive organizations, the focus will be on knowledge workers—that is, people who need to acquire new knowledge to perform their jobs. *[All references to the competence lemon can be found in Chapter 1, references to the competence loop can be found in Chapter 2, and references to REPI can be found in Chapter 4.]*

Sometimes project portfolio management is part of the PMO's responsibility, and sometimes it is managed elsewhere within the organization. There are also organizations that do not work with project portfolio management in a structured way at all. However, to have a structured way of managing the strategic and prioritized projects in the organization, it is preferable to work with project portfolio management.

---

* Reflection, Elaboration, Participation and/or Practice, Investigation

Several of the factors in the utilization mechanism of the competence loop are closely related to project portfolio management, especially because this is one way of meeting strategic goals in terms of prioritizing projects and competence. The latter is one main purpose of the utilization mechanism in the competence loop, although this is part of the interaction between the parent organization and projects. To be able to be successful with the different projects in the project portfolio, the organizational leaders need to appoint the right project managers for the different projects.

To appoint the right project manager, which in many cases is a part of the PMO's responsibilities, will be our next focus.

6.1 How to Appoint the Right Project Manager

How many times have we seen a project managed by the wrong project manager? There is a belief that if you are a good project manager in one context, you will automatically be able to manage any project. Is this really true? In Chapter 1, the opposite example was presented, and this example will be further elaborated in this chapter.

In that example, a project manager who had successfully managed projects in one area was not the right choice in another, adjacent area. He was one of the project manager heroes in the organization and had managed several successful projects within the area of a specific system solution. Furthermore, he was respected by the project team and in other parts of the organization and was considered a senior project manager—also mentioned as a potential manager in the organization.

The project manager had started his career as a young system developer within the subject area, continued to be a system architect, and later on became a project manager. The adjacent area was in crisis and was in need of a skilled and experienced project manager, for which reason the management team appointed this person as project manager for the critical business project in the adjacent area. But it went wrong. The project was far below expectations in terms of cost, time, and delivery. The project manager failed to manage the project and moved back to the earlier area, where he continued to manage successful projects. The management team had only considered the previous history when appointing him as project manager in the adjacent area. What was the right approach in this example?

Before appointing a project manager for the example above, we shall look into how we can use the competence lemon as a tool for appointing a project manager. The competence lemon says that context, organizational culture, and identity impact the application of competence. In selecting a project manager, context can be many different things and can be divided into two dimensions: the *general context* dimension and the *project characteristics* dimension.

6.2 General Context Dimension

6.2.1 Industry

The first general context dimension is *industry*. In which type of industry are we acting: construction, oil and gas, IT, public sector, retail, supply chain, etc.? Different industries have different prerequisites for which the project manager needs to act in different ways. Different industries also develop specific cultures that the project manager needs to consider. In addition, previous knowledge and experience within the subject area facilitates the ability to manage the project. For example, as mentioned above, it is difficult to manage an oil and gas project without any previous knowledge and experience in that particular industry.

6.2.2 Globality

The next general context is *globality*. Is the organization acting globally or in a limited market? This has an impact on the degree of cultural communication skills the project manager needs to have. A global organization can, in many cases, have a more complex stakeholder situation, with different stakeholders in different parts of the organization. The ability to manage complexity, good leadership qualities, and social capability will facilitate the management of complex stakeholder situations. This dimension also includes heterogeneity in culture. In a heterogeneous cultural context, the project manager needs the ability to establish a way to communicate across cultural borders.

6.2.3 Organization

Another dimension is *organization,* which describes the type of organization in which the project will be conducted. Is it an R&D organization, within supply chain, or another type of organization? The difference between this dimension and the industry dimension is that in some cases the industry can be public sector and the organization IT, for example, while in others the industry can be IT. In the first example, the IT organization is in most cases a non-core organization, while in the latter it is the core business.

6.2.4 Size of the Company and the Organization

The next general context dimension to consider is the *size of the company and the organization.* The size of the organization and the company matters. It can be argued that these are two dimensions: here we will manage them as one but in reality consider them as two. In a large organization, there are normally many processes, policies, and routines to consider and manage. This can mean greater efforts for administration, greater stakeholder management, etc. On the other side of the coin, if the organization is a small IT department in a large supply chain company, the processes, policies, and routines may not be adapted to the IT department's specific needs.

6.2.5 Market Maturity

Market maturity is another dimension of context and refers to the maturity of the market in which the company acts. In an immature market, a project manager generally needs to handle more unknowns and uncertainties than in a project in a more mature market. The level of market maturity impacts not only the project manager's ability to manage complexity, but also his or her personal and social capabilities, as well as leadership qualities.

6.2.6 Organizational Culture

The *organizational culture* is the next dimension of general context. The cases described in Chapter 3 show the impact of organizational culture on project management in general. In a positive and functional culture, trust and transparency are generally a part of the culture, and the project manager has a cultural baseline to start with. In a dysfunctional *organizational* culture, the project manager needs to have good or excellent communication skills and personal capabilities to establish a positive *project* culture.

6.2.7 Knowledge Intensity

In the context in which the project is carried out, different levels of *knowledge intensity* can occur. One example is an R&D organization that has to deal with a high level of new product ideas and innovation, compared to a supply chain organization that is based on a repetitive way of working. In the first example, the project manager needs to have personal capabilities to encourage innovation and new ways of thinking, whereas in the latter example, the focus may be on minor improvements to established ways of working.

6.2.8 Level of Specialist Orientation

Level of specialist orientation is the next dimension of general context. As outlined in Chapter 3, specialists have deep knowledge in their specific subject areas. Specialists or experts have a tendency to fly in, do their job, and fly out.

The consequences of this behavior are that the organization does not learn from these experts, and that the experts do not learn from the actual project in which they participate. The level of specialism impacts which skills the project manager needs. In a team of generalists or in a self-organizing Scrum team, the project manager does not need a high level of ability to manage heterogeneity, which is a part of the ability to manage complexity from the competence lemon.

6.2.9 Project Management Maturity

Project management maturity is another general context dimension to consider. Different organizations have different levels of maturity in their way of working with projects. Many organizations have well-defined project methodologies, while others have more or less nothing. A project manager used to acting in the context of mature project management could encounter difficulties managing a project in an organization that does not have a proper way of working with the projects in place. In the context of an organization with immature project management, the project manager needs to be more flexible and to have a better understanding of how project team members are acting, especially when the project team members do not always understand that they are working on a project at all. In an organization with immature project management, the steering groups also tend to behave in an immature way, which impacts the project manager's way of managing projects.

6.2.10 Previous Change History

The last general context dimension is *previous change history*. This perspective of context refers to the history of the organization. Does the organization have a history of major changes, and how have those changes been carried out? Why does this have an impact on the appointment of the right project manager? Most projects contain an element of change management: the nature of a project is to move from situation X to situation Y with some degree of uncertainty and unknownness. The people in one organization can be tired of change, while in another organization people can be positive about moving to something new and exciting. The history of the organization will have an impact on the skills the project manager needs to have.

Table 6.1 summarizes the ten context dimensions to consider when appointing a project manager to a specific project. What is important to highlight is that the project manager does not need to excel in all dimensions, but rather the project manager has different levels of strength in different dimensions of the competence lemon. The philosophy of the competence lemon is not to classify people; it is a way to map competence to context and support the development of people to manage new and unknown situations.

Table 6.1 General Context Dimensions

No.	General Context Dimension
1	Industry
2	Globality
3	Organization
4	Size of the company and the organization
5	Market maturity
6	Organizational culture
7	Knowledge intensity
8	Level of specialist orientation
9	Project maturity
10	Previous change history and organizational stability

The other dimension of context is project characteristics. Whereas general context could be seen as the context in which the project will be managed and has an outside-in perspective, project characteristics form the project and describe it from the inside out.

6.3 Project Characteristics Dimension

6.3.1 Project Management Approach

The first dimension that relates to project characteristics is the *project management approach*. In a traditional waterfall-based project, the project manager will focus on following the project plan and avoiding change. If any change requirement occurs, a change request leading to changed plans needs to be analyzed and decided on. In agile project management, change is natural, and prioritization is done based on added value. The different project management approaches require different project management skills.

6.3.2 Type of Project

Another project characteristics dimension is *type of project*. Is the project within finance, supply chain, procurement, HR, production, R&D, or other area in the company? Different functional areas have different cultures, history, project maturity, processes, etc. Does the project manager have previous knowledge and experience in the area?

Another aspect is the competence level of the people working on the projects, which can differ between different functional areas. Managing an R&D project differs from managing a project in the production line. The project manager needs different skill sets for these two project types.

The project can also be an organizational change project, which is generally different from other types of project. Most often, they have few fully allocated team members, and the project affects many people in the organization. This can be in terms of new business processes or organizational change. The level of communication is normally higher than in most other projects. An organizational change project requires high leadership qualities and an ability to manage complexity. Also, a high degree of social capability supports the project manager in managing this kind of project.

6.3.3 Time

The next dimension of project characteristics is *time*. Time has a significant impact on how to manage a project. Longer projects normally have a higher degree of uncertainty, and the probability that the scope will change is much higher. The project manager will need to manage change to a higher degree than in a shorter project. Another view of time is that the probability of changing project team members will increase, which requires a higher focus on competence allocation and onboarding of new team members, as well as on working on continuous team building.

6.3.4 Size

Size is a dimension of project characteristics and can be viewed from different perspectives. One perspective is people—in many projects team members may be divided into different teams managed by sub-project managers. In this case, the project manager is leading leaders, which requires different leadership qualities from leading a smaller project team. The other perspective of size is money, as in having a large project budget. In this case, the project probably has many suppliers, and the project manager needs to have leadership qualities to lead a temporary organization that interacts with one or many external parties.

6.3.5 Task

The next project characteristics dimension is *task*. This dimension refers to the actual task that the project will work on—for example, incremental development of a product, developing a training program, decommissioning, putting a man on the Moon, rolling out a recruitment solution, etc. Different tasks require different project management skills. If the task is development of a product using internal resources, for example, the project manager will have a team focus and will ensure that the product meets requirements and expectations. On the other hand, if the project task is to implement a solution in an operational business, the project manager needs to have more focus on stakeholder management and communication and to work hard on the actual implementation plan. These two examples require two different leadership styles and project manager competence profiles.

6.3.6 Level of Uncertainty

Level of uncertainty is also a project characteristics dimension. Does the project have many unknowns and uncertainties? A project can often be difficult to start up because of circumstances such as a lack of resources, unclear scope, funding problems, etc. One reason can be the level of uncertainty, such as unclear goals, new technology, never having done it before, or any other uncertainty. Furthermore, there can be many uncertainties to manage during the project—for example, what happens if the new technology does not work? The project manager for this kind of project needs to have strong personal capabilities, high ability to manage complexity, and good leadership qualities. He or she also needs to be good at managing risks and managing key stakeholders in order to be successful in the role.

6.3.7 Level of Complexity

Level of complexity also forms part of the project characteristics dimension. Complexity can have different angles, as follows.

- **Supplier complexity.** The first angle is *supplier complexity*—for example, many suppliers, suppliers from different countries, or suppliers that have an unstable financial situation.
- **Heterogeneity of knowledge.** The second angle of project complexity is that a project may cover several functional areas, which means that a range of different knowledge areas and subject expertise need to be considered. Where there is *heterogeneity of knowledge,* the project manager needs to have skills in establishing a positive project culture in which people respect each other's knowledge and collaborate to achieve common goals.
- **Cultural complexity.** The third view of complexity is *cultural complexity,* meaning that team members or suppliers come from different cultures. A multicultural project requires the project manager to have some level of intercultural communication skills or support in intercultural communication.

- **Stakeholder complexity.** The fourth angle of complexity is *stakeholder complexity*. Stakeholder management is an important area for a project manager. The level of complexity in this area can differ. If the project is highly prioritized, it is probably under the lens of executives in the company. Good communication skills are essential in this situation. Another situation is a project that is not wanted by some parts of the organization, wherein the project manager needs to develop a tactical plan to manage those stakeholders. In all large organizations, and in smaller ones too, there is a degree of politics, in which different people have different agendas. Managing a project in an organization with a high degree of politics requires a good ability to manage complexity, personal capabilities in terms of self-confidence, diplomacy, etc., and good leadership qualities. In addition, social capability is needed to manage stakeholders on different levels. Furthermore, stakeholder complexity can also mean that there are many stakeholders—for example, in an infrastructure project in a city that impacts everyone's daily life. Conversely, if the level of stakeholder complexity is low, a project manager with lower competence in the above-mentioned areas can be appointed. There are other views of complexity that need to be considered case by case.

Table 6.2 summarizes the seven project characteristics dimensions to consider when appointing a project manager to a specific project. As with the general context dimensions, it is important to highlight that the project manager does not need to excel in managing projects with all types of project characteristics. The reason for outlining the project characteristics is to match project manager competences with project characteristics and analyze potential gaps.

The seven dimensions form the project characteristics and, together with the general context, are the bases for the project. We have analyzed the context in which the project shall be executed and outlined the project characteristics. Now we go on to evaluate the project manager for the project.

Table 6.2 Project Characteristics Dimensions

No.	Project characteristics
1	Project management approach
2	Type of project
3	Time
4	Size
5	Task
6	Level of uncertainty
7	Level of complexity

6.4 Select the Right Project Manager

This section outlines the steps for selecting the right project manager for the project. We will return to the example of selecting the wrong project manager from the beginning of the chapter and see how we can use the new approach in that example and look at what went wrong.

First, we need to use the general context and the project characteristics described above to form a baseline for the need for a project manager. In this step, all dimensions are evaluated, which will result in a project manager competence profile as outlined in Figure 6.1.

The competence profile considers all dimensions in the competence lemon and the context in which the project will be managed. This is done by evaluating the dimensions in the competence lemon and putting them into the project context.

The next step is to establish the competence profile of a proposed project manager considering all dimensions in the competence lemon as well as the project context, as exemplified in Figure 6.2.

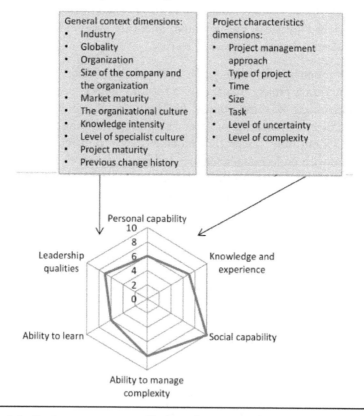

Figure 6.1 Project manager profile based on general context and project characteristics dimensions.

Figure 6.2   The proposed project manager's competence profile.

We now have the required competence profile and the proposed project manager competence profile. In the next step, we compare the two competence profiles to analyze potential gaps, as outlined in Figure 6.3.

In Figure 6.3 we can see that the proposed project manager has a higher degree of personal capability and ability to learn than the required profile. However, the proposed project manager for this specific project shows a slightly lower degree of leadership qualities, ability to manage complexity, and knowledge and experience. The social capability is high, as the project requires. What can we do with this information?

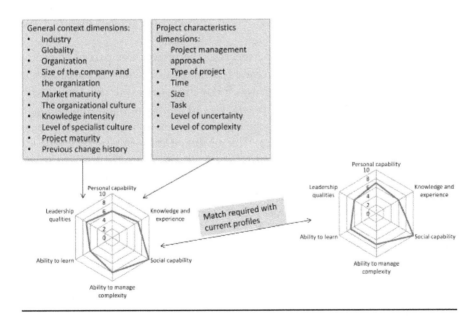

Figure 6.3   Match required competence profile and proposed competence profile.

We have different options. One option is to say that the proposed project manager is not competent enough to manage this project; another is to increase the project manager's competence in appropriate areas through some training exercises. His or her ability to learn is rather high. It is also possible to add a person who can support in leadership and manage complexity, and in this way support the project manager to manage the project. As seen, there are different options based on what we know about the pre-requisites and the proposed project manager's competence, considering all dimensions of the competence lemon. The important thing is that we are aware of the gap and can utilize this knowledge to outline different options.

Let's return to the example at the beginning of this section about the project manager who did not succeed when moved to another solution area. The important thing is that we are aware of the gap and can utilize this knowledge to outline different options

Looking at the general context, the company was acting in the telecoms market. It was global, and the organization was delivering support tools and processes for the company's internal use. The organization consisted of about 200 employees in a company that in total had more than 100,000 employees. The market was somewhat mature, and the organizational culture was considered strong, functional, and with high knowledge intensity, with many engineers in the organization. Furthermore, the solution area could be considered as having high heterogeneity in terms of knowledge and culture, with two suppliers from different parts of the world. Moreover, the solution area consisted of a mix of some specialists and several generalists. In addition, the project maturity was very high.

From a project characteristics perspective, the project was part of the company's customer support, having some IT components, a non-core area for the company. The project methodology was following a strict waterfall model with a well-working tollgate model. The task was to deliver new IT solutions and updated documentation, to train people in the new solutions, and to implement new business processes for the area. The solution area normally delivered two releases per year, for which reason every project lasted about six months.

The size of the project could be considered rather small. The team consisted of around 10 people, excluding the suppliers, comprising solution architects, requirement handlers, testers, process developers, and trainers. The team culture was tough and direct and valued knowledge. The level of uncertainty was low, but on the other hand, a certain level of complexity existed, because requirements should be gathered in different organizations and synchronized with the two suppliers. Stakeholder complexity was high: since many organizations used the solutions, funding was coming from different sponsors, and the selection of suppliers was questioned within the company.

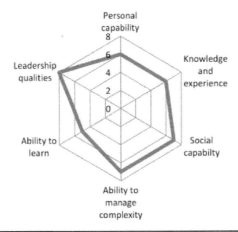

Figure 6.4   Required project manager profile.

The required competence profile for the project manager can be seen summarized in Figure 6.4. Because the organization was mature in the market, it had established ways of working, high maturity in project management, etc.; no extreme competence levels were required. However, the stakeholder complexity was rather high, so the required social capabilities, the ability to manage complexity, and leadership qualities were higher than average. Due to the special team culture, in which knowledge was valued, communication was direct, and the atmosphere was a bit tough, the required leadership qualities were rather high in order to be able to manage the team and the suppliers successfully. Analyzing the project manager, we got a totally different competence profile, as outlined in Figure 6.5.

Figure 6.5   Actual project manager profile.

Personal capability was not much below what was required. The project manager's ability to learn was high. He had started as a system developer, continued to be a systems architect, and ended up acting as project manager in the same area. But the learning in this case was knowledge in a specific subject area: he was managing his previous projects based on deep knowledge in the solution and in the subject area. In addition, he was highly respected in his previous team because of his deep knowledge and experience.

In one of the critical dimensions of competence for the new project—leadership qualities—his level was low: in other words, he did not possess any particular leadership qualities at all. Instead, he managed his previous projects based on knowledge. His ability to manage complexity was lower than average and far below what the new project required. In his previous roles, he had gained new knowledge step by step.

The gap between the required competence profile and the project manager's profile was significant, leading to a situation in which he could not manage the project. Without leadership qualities, he could not lead the team, which operated in a tough climate, had direct communication, and valued knowledge. The solution knowledge that the project manager had from the other solution area was not valued in the new solution area. In addition, the project manager failed to manage the complex stakeholder situation and could neither lead the synchronization of requirements between organizations nor secure funding from key stakeholders.

The project manager moved back to his previous solution area and could continue to use his knowledge in that area. His ability to learn broadened his knowledge in that area even further, and he continued to grow from there. The step to take on a new project in another area, even if it was in an adjacent area, was too great. The right career path for the project manager would be more in a specialist role than in managing projects.

In this example, the management team appointed the project manager based on his reputation in the current solution area without considering why he was successful. It is possible to act and lead based on knowledge, but in that case you are limited to lead in the area in which you have that knowledge. Leadership qualities are also about giving others information in a way that they can conduct their work, which is the case when leading by knowledge. The risk with leading by knowledge is that you become better than your team members and, in the long term, this hinders people's growth. The learning from this example is that all dimensions of the competence lemon have to be considered, and that they should be considered in relation to the context in which the project will be managed.

The measurement of different competence dimensions can be done in different ways. Many organizations measure personal and social capabilities as well as leadership qualities. Furthermore, knowledge is evaluated in relation to performance management. The two new dimensions of competence—ability to manage complexity and ability to learn—are not fully supported in many traditional measuring tools but can partly be found in tools such as Lominger (and others). It is also possible to measure in a qualitative way, estimating the levels. One of the recommended ways—which will be presented in Section 6.6, about agile performance management—is to discuss competence based on the six dimensions in the competence lemon between manager and co-worker, and in this way make the measurement less formal and quantified.

Having discussed how the competence lemon can be used to select the right project manager, we will continue on the same theme and see how it can be used for competence development.

6.5 The Competence Lemon as a Tool for Competence Development

The example in the previous section touched on how the competence lemon could be a tool for competence development. The project manager maybe should not be project manager, but should instead continue to develop knowledge within a subject area and, based on previous knowledge, move on to another area, being more of a specialist. The ability to learn was high, and he had taken steps within the previous subject area. One reason for his moving to the wrong position was that not all dimensions of competence in the competence lemon were considered.

In a knowledge-intensive environment, one's previous knowledge is of less importance than in an environment that has a low degree of knowledge intensity. Instead, the other dimensions of competence will assume higher importance. This can be seen as a contradiction, as knowledge intensity is based on renewal of knowledge. The word *renewal* is key: it is the other dimensions that support renewal of knowledge. Of course, a high degree of ability to learn facilitates acquiring new knowledge, but one's social capability also supports in acquiring new knowledge, as knowledge sharing is a competence-generating activity.

Competence development is not only about acquiring knowledge, but is also about training your skills in such a way that your application of knowledge and experience becomes efficient, which can be measured in terms of performance. It is possible to train personal and social capability by practice. Moreover, it is possible to train the ability to manage complexity by being exposed to complex situations such as stakeholder interactions, linking information from different domains into new knowledge, and much more.

In a knowledge-intensive, project-intensive context, project team members' competence development should be linked to projects in an arena with transparency between competence development planning and project planning. Several factors enable the generation of new competences in projects.

- Knowledge sharing between team members
- Providing an environment in which team members feel confident and have trust in the team
- Communication between team members and with people outside the project, including external sources
- Problem solving and experimentation
- Cross-functional learning between different resources with different competences
- Visibility and transparency of information in the project and in the organization
- On-the-job training, during which more experienced resources train those who are less experienced
- Ensuring that people feel motivated and have a positive attitude to work
- Ensuring that people have the ability to work with challenges and try new ways of working
- The ability to absorb external competences

This book presents two frameworks and methodologies supporting competence development. The first is the competence lemon, in which we can set goals for the different dimensions. The second is REPI, which supports the actual competence development activities.

6.5.1 The Competence Lemon

The competence lemon is an excellent framework for setting goals and for discussion regarding current competence levels. As we saw in the section about appointing the right project manager, it can also be used to define the competence requirements for a role in a specific context.

Let's take an example of a newly graduated engineer. She is a very young, social person but suffers from low self-confidence and needs to develop some personal skills. In her job, she needs to acquire more knowledge in her subject area and to improve her ability to manage complexity, because working as an engineer requires that you can manage information from different domains and link it to new solutions. Currently, the young engineer does not need to develop her leadership qualities, and the manager and the engineer decide that her current level of ability to learn is sufficient. The manager and the young engineer agree on the current situation and the goal setting according to Figure 6.6.

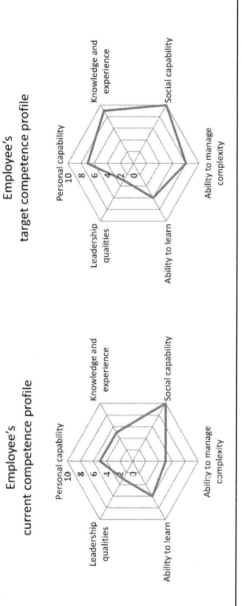

Figure 6.6 Employee's current and target competence profiles.

We can see here the goals and the plan regarding the areas in which the young engineer is going to develop. A project is effectively a learning arena, although it always consists of some uncertainties and unknowns, which means that there will be some kind of problem-solving activity. Just working on projects will increase a person's knowledge base and maybe also the other dimensions in the competence lemon, but an active competence development plan needs to have more components to be efficient. One excellent methodology for competence development is REPI, which is described in detail in Chapter 4.

6.5.2 How Can REPI Be Used in the Example of the Young Engineer?

In the project, the young engineer gets some problems to solve. But instead of just solving the problems, the project work is linked to the REPI modes. She will be assigned a coach who will lead her through the different steps in the competence development plan. The coach is a more experienced engineer with a suitable competence lemon profile. First the young engineer has to elaborate on different ways of solving the problem. She needs to seek information from different sources in relation to elaborating on the problem. To get feedback and to practice her presentation skills (part of the personal capability), she presents different proposals for the team.

After the presentation, she spends some time reflecting on both her proposals and how the team reacted to the solutions. She also has REPI meetings with her coach to work with the different REPI modes—reflection, elaboration, practicing, and information gathering (investigation). The REPI meeting ends with new REPI activities as the next step. In the REPI activities, she acquires more knowledge within the subject area by looking at the problem from different perspectives; she trains her ability to manage complexity by linking information from different sources; and she trains her personal capability by practicing presentation and managing team meetings.

The young engineer also gets other REPI tasks—for example, conducting training for others to acquire more knowledge and experience, but also to improve her personal capability.

By connecting the dimensions of the competence lemon with the different REPI modes, competence development is built into daily work, and the perfect arena for these activities is the projects in which people participate. One important factor in using this way of developing competence is that the whole organization understands that competence evolves through projects and that the projects are part of a permanent context in which project managers also have other duties than just managing the project.

In a project environment, the accumulation mechanism from the competence loop is managed in the project. If the project manager is aware of the factors in the mechanism and of the factors that have an impact on generating new competence, he or she can make the project team members more efficient by enabling competence-generating activities and establishing the project as a competence arena. By implementing this way of continuous competence development, the organization's innovative capabilities will increase.

A knowledge worker learns through working and trying new and unknown things. The learning process is a continuous process that sometimes has a low level of learning and sometimes higher. With continuous learning, we also need continuous follow-up of what people learn. In the next section, we will discuss how performance management leaves the yearly appraisal to an agile process.

6.6 Agile Performance Management

The assimilation mechanism in the competence loop describes how people in the organization understand and interpret what kind of competences the co-workers develop by working on projects or other activities. A part of the assimilation mechanism is performance management, which is the process that ensures that business goals are being met in the most effective manner and is built on communication between the line manager and the co-worker. In a knowledge-intensive context, traditional performance management processes tend to be inefficient, especially as markets move faster and faster. The performance management process needs to be more agile and adapt to change.

6.6.1 Measuring Performance

Competence and performance can be measured in different ways. When knowledge constantly needs to be renewed, the actual measurement of knowledge is of less importance. As humans, we strive to categorize and put labels on people "he is a four out of five" and "she is only good at procurement." There is a danger in grading people, even if it is an easy way of categorizing. Most often you and the one that grades you do not have the same opinion, which will lead to tensions. Generally, we are aware of the different opinions of grading, where the line manager and the employee have different opinions, for which reason the tensions tend to be permanent.

Worse is when "below expectations," "meets expectations," or "exceeds expectations" are used for measuring performance. If a person is a high performer, the expectation will be high, and "meets expectation" will be a high performance; for someone else, "exceeds expectations" may not be as good because the original expectation was low. The expectation levels only demotivate and generate tensions. Worst of all is when expectation levels are connected to the salary process. A high performer deserves a good salary, even if he or she only meets expectations. Exceeding expectations is a result of an extraordinary contribution and could perhaps be rewarded just one time and not form a part of the yearly rise in salary.

Because of that, this book is about competence management and what generates new competence; we will look at performance management from another perspective—one in which the process supports the generation of new competence and does not demotivate or create tensions.

The traditional way of managing performance is the yearly appraisal with a follow-up talk after six months. In the appraisal, goals are set for the year and will, in the best cases, be linked to the organizational strategy and goals. In general, people feel that it is difficult to set goals that are related to the business goals. This feeling is valid for both managers and co-workers. The issue today is that everything moves so fast, leading to most goals being obsolete after a year. The goals have to be changed or new ones have to be added, leading to a situation in which the process will be inefficient and the talks more focused on what goals to set and less on what people have learned and how the new competence can be used.

The other issue with the traditional appraisal process is that knowledge workers tend to work on many different projects and activities, which makes it difficult for the manager to follow up on performance. Some organizations try to use 360-degree evaluation with feedback from different parties in the organization. This kind of evaluation also tends to be inefficient and unfair. Because the context impacts how competence is applied, feedback from many parties can be difficult to interpret. If a project manager makes decisions that are necessary but which affect project team members in some negative way, the feedback could be negative, even if the decision was totally right to achieve the business goals.

There are many aspects to evaluating, and the weakness of this kind of feedback is that the person giving feedback might only interact with the other person in specific situations. One example is that that the steering group members only interact with the project manager in steering group meetings, and not in other situations or as part of the project context.

6.6.2 An Agile Approach to Measure Performance

Using an agile approach to interpreting and understanding what co-workers have learned will move performance management from categorizing, meeting expectations, and tensions to talks during which different dimensions of the

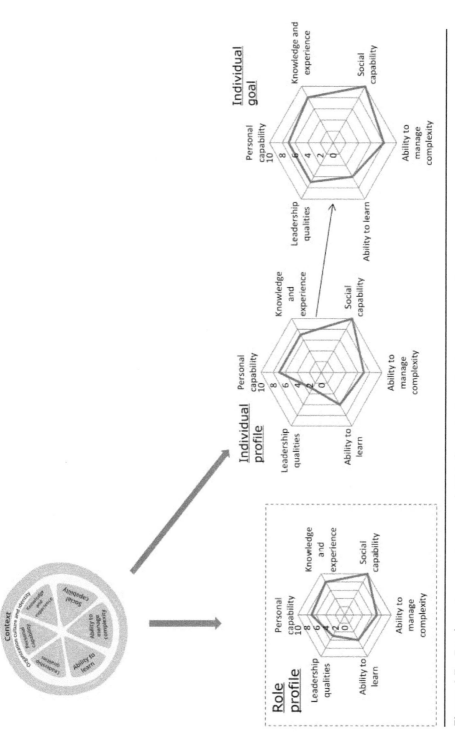

Figure 6.7 Employee's competence goal. [*This figure is described on next page.*]

competence lemon, as well as achievements, are discussed. The talk has to be continuous, which makes it agile and adaptable to the current situation. It is possible to keep goal setting and a yearly/half-yearly follow up, but the continuous agile talk is the basis to follow up on achievements, learning, and the application of competence. Continuous talk will be more informal and will focus on continuous improvement rather than big steps.

The talk can be performed using REPI meetings as the basis. The advantage of the REPI meeting is that it focuses on competence and performance. If the line manager acts as the REPI facilitator and the employee as client, the meeting will follow up on performance and competence development. The frequency of the meetings can differ from case to case, but they should not be held less than every fortnight, and meetings are preferably short.

In the section describing the appointment of a new project manager, we showed how the role profile could be mapped toward the individual profile of the proposed project manager. In agile performance management, this will be taken one step further, to include the individual's competence goal, as illustrated in Figure 6.7.

The current role is determined based on the competence lemon as well as the co-worker's current competence profile. However, the co-worker also has a long-term goal that requires additional competence development, or it could be part of the succession plan. In the example in Figure 6.7, the current role does not require strong leadership qualities, but the long-term plan for the co-worker is to move on to a leadership role that requires a higher degree of leadership qualities and the ability to manage complexity. In the agile performance management approach, the goal for the current position as well as the long-term goal are considered and followed up using the competence lemon as a basis for competence development and follow-up talks. Preferably, REPI is used as the methodology to acquire the new competence.

6.6.3 Agile Performance Management in a Project-Intensive Context

In a project-intensive environment, people are working on one or several projects, managed by different project managers. In general, the project manager does not pay attention to project team members' competence development or what they learn in the project. Their focus lies on the project outcome and delivering according to plan.

However, the project manager has a responsibility in the organization in which he or she acts, and the project should also have responsibility for some parts of competence generation. If so, the project manager also should play a natural part in assessing project team members' development. In a project-intensive environment, the line manager and the project manager need to have continuous follow-up on project team members' competence development and what they have learned in the project. The line manager is responsible for this coordination and synchronizes it with continuous talks with co-workers. The frequency can vary, meaning that the follow up between the line manager and the project manager does not need to be as frequent as the talks between the line manager and the co-worker.

Another activity that can be included in projects is frequent lessons-learned or retrospective meetings. These meetings should include follow up on what people have learned in the projects and how this can be used in the future—for example, if there are new innovation opportunities or things that could be done differently. Today's lessons-learned and retrospective meetings working with Scrum are focused on the way of working and how working processes can be improved. Follow up on working processes is good, but innovative ideas can get lost and organizational learning does not work as well as it could. With specific follow up on learning sessions in the projects, the organizational learning will improve and lead to more innovation, both incremental and radical.

One group that is often forgotten in traditional performance management is the external workforce, which can include contractors, consultants, suppliers, or other non-employed knowledge workers. Contractors and consultants are often very important resources in projects and in the organization as a whole. Managers and other key persons in the organization need also to follow up on their learning and how their competence can benefit the organization. Here, competence-focused lessons-learned and retrospective meetings are important activities to absorb external knowledge, which will also improve organizational learning. This kind of activity will also assess all dimensions of external resources' competence and be used to understand how those resources can be allocated to projects and other activities later on. The capacity to learn from external parties—the organization's absorptive capacity—is crucial for an organization's survival and for the organization to be competitive in the market.

6.6.4 Summary of Agile Performance Management

In summary, we can conclude that performance management has to move from the yearly cycle to be more agile and adaptive. The way to implement agile performance management is to use a more informal but continuous competence development talk based on the different dimensions in the competence lemon.

The other part of agile performance management is using competence-focused lessons-learned and retrospective meetings on a frequent basis to follow up on what people have learned during the project. Both activities need to include external resources to improve the organization's absorptive capacity. In addition, REPI meetings are preferable for the dialogue between the line manager and the employee, as well as being a methodology to follow up on team performance.

In this section, we have discussed how to understand and interpret newly generated competences by different kinds of formal and informal processes that are part of the assimilation mechanism in the competence loop. In the next section, we will elaborate on the concept of core competences and how the organization can work with those in order to be competitive in the market and enable innovative capacity in the organization.

6.7 Identifying Core Competences

Core competences are those capabilities that are critical to a business achieving a competitive advantage. They are built on continuous improvements and enhancements (Eden and Ackermann 2010; Hamel and Prahalad 1994) and are manifested in business processes and activities (Agha, Alrubaiee, and Jamhour 2011). Core competence is a significant determinant of organizational performance and competitive advantage, in the sense that more competence leads to a higher degree of organizational performance and competitive advantage. Furthermore, responsiveness and flexibility can be seen as two dimensions of competitive advantage, and these are antecedents to organizational performance.

Individual knowledge (part of human capital) becomes institutionalized and codified (organizational capital) and is transferred between people through networks (social capital) to form an organization's intellectual capital (Subramaniam and Youndt 2005). This means that knowledge can be codified and stored, whereas competence is a part of the people working in the organization. The core competences are those that the organization needs to reach sustainable competitiveness, which means that they have to be aligned with the organization's strategic goals.

Sometimes the concepts of core competence and key competence are mixed. They are more or less the same, but have a slight difference. *Core* competences are, as described above, those capabilities that are critical to a business's achieving a competitive advantage, whereas *key* competences are those that we need to achieve high performance and carry out necessary tasks. Key competences can be expert skills or knowledge in different areas that are not the core business. The organization might need resources with key competence in support processes that are not a part of core business—for example, how to manage logistics to reduce costs or drive the recruitment process in some countries. Those resources could possess knowledge and competence that is key competence but not core competence.

In an organization, we have many different categories of competence. Figure 6.8 visualizes the different kinds of competences that we need or which occur in an organization.

In an organization, people need to possess competences that are common to all, such as how to report time, understanding corporate policies and procedures, how the corporate code of conduct works, being able to write documents, and other similar knowledge that is the foundation for the organization. Based on their experience, employees also have different competences not considered as core competences, but which are needed to perform different tasks. Usually, the organization has expert knowledge that not is a part of core competences. This might be knowledge on taxes, customs, HR processes, etc. Normally, the organization also uses an external workforce that possesses different kinds of competence necessary to perform different tasks.

As market conditions change, people in an organization also possess knowledge and competence, such as about old technology or about products that are no longer needed. People can also have different competences not needed in an organization but valuable on the market, and these can be considered competence risks. For example, if a person possesses knowledge in a specific technology that is not currently used in the organization, there is a risk that that person will leave the company. This situation can also be a possibility: Knowledge in a specific technology might be used to develop new products or services. Another situation regarding competence is that the organization does not have the competences that it needs. In this situation, organizational leaders need to develop a strategy to acquire the competence, perhaps external sourcing, recruitment, or competence development.

Finally, the organization has its core competences. How can an organization identify its core competences?

To analyze and identify an organization's core competences, it is necessary to start with the strategy and define the business goals for the long- and mid-term. These goals will be the foundation for what the organization is going to achieve. The next step is to describe the organization's potential successes and failures in achieving those goals. In this step, it is necessary to consider not only technical competences but also business-oriented competence—for example, how to enter new markets. Core competences in entering new markets might be intercultural communication skills, establishing partnerships with external partners, and similar competences. A pitfall is to focus too much on technical knowledge as a core competence and forget knowledge about different processes, or not consider all dimensions of the competence loop.

When describing potential failures to reach goals, it could be that the organization does not have the ability to adapt to fast-moving market conditions. The organization has all the technical knowledge, but it is not agile enough to move with or even be in advance of the market. This kind of market is most often knowledge intensive and, in knowledge-intensive organizations, learning is a core competence.

When core competences are identified, the competence gap in the organization needs to be analyzed and defined. The competence gap is the difference between the estimated current competence and the core competences, and analyzing and defining it can in many cases be the most time-consuming step in the process. The competence gap is the basis for strategic decisions on how to fill the gap, which are mainly composed of four different options:

1. Recruit people with the right competence
2. Develop co-workers within the organization
3. Source externally
4. Acquire a company with the competence

Different gap-filling strategies can be used for different areas. Normally, different combinations of these options are used to fill the gaps.

The concept of core competence is crucial to managing competence in an effective way. Many of the factors in the competence loop rely on the classification of competences as in Figure 6.8. This can be done in entities within an organization and preferably will be linked to the company's strategy in a top-down process.

However, it is possible to identify core competences in an organizational entity within a company with a bottom-up approach. This is accomplished when

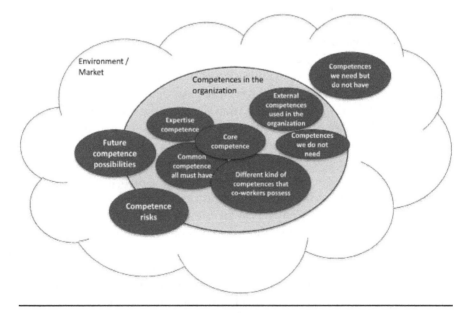

Figure 6.8   Competences within an organization.

the organizational entity within the company defines its goals and strategies and, based on these goals, identifies the core competences needed to achieve the goals following the steps outlined above. In many cases, it is easier to start with a smaller organizational entity, because people in the smaller organization can understand what competences are core to reaching their specific goals.

However, the recommendation is to start with a top-down approach to map company strategies to the smaller organizational entity's goals and then continue with the bottom-up approach to identify core competences needed to reach these goals. In the subsequent step, the core competences defined by the smaller organization needs to be consolidated to avoid duplicates, incorrectly defined core competences, and sub-optimization.

The concept of core competence is part of the transformation mechanism in the competence loop and is the basis for the next mechanism, utilization. It is also a platform for interpreting newly generated competences in the assimilation mechanism in the competence loop.

In the next section, we will move the focus to the HR perspective of managing competence.

6.8 The HR Perspective

Do not look at what your co-workers can do for you today;
think about what they can do tomorrow.

— CEO for Swedish branch of an executive network

In most companies, HR has overall responsibility for the process of competence classification and development. How can HR benefit from the principles, approaches, frameworks, and tools presented in this book? This section will take the HR perspective and look into how HR can support the organization in managing competence in line with the organizational goals.

The competence lemon can be used in different ways from an HR perspective. One way is as a support for talent management. Talent management can be seen as an organizational commitment to recruit, retain, and develop talented employees to achieve a sustainable competitive advantage: HR should be the drivers of this process by ensuring that managers identify talented employees.

6.8.1 Identifying Talented Employees

But what is a talented employee? Many companies consider all employees to be talented, whereas other companies only track the talents of people suitable for executive levels. Using the competence lemon approach, every employee is a talent, especially because it is possible to develop all dimensions of the competence lemon based on the need for a specific role. Furthermore, this view of talent is very useful for succession planning, in which the organization aims to have a successor for a specific role.

Specificity

Normally HR works by defining job profiles for *specific positions*. The main difference between a job description and the competence lemon is that the competence lemon takes a holistic perspective on competence, taking into account the context, the organizational culture, and the identity, and how these impact the application of competence. A standard job profile does not consider the context, the organizational culture, and the identity.

An example is the position of warehouse manager in a large company. The standard job profile is the same for all warehouse managers, independent of the context or whether the culture is different between different parts of the organization. The competence lemon takes into account the contextual perspective and says that the application of competence differs between different environments.

Furthermore, the competence lemon looks at different dimensions of competence, acknowledging that knowledge or experience is only one dimension, whereas most job profiles only cover knowledge and experience, some personal capabilities, leadership qualities, and some social capabilities.

Internal Mobility

One major part of organizational learning is *internal mobility*, in which HR should again be the facilitator, as this is considered an important knowledge-sharing activity and a way to develop and retain employees. Internal mobility is closely related to talent management processes, especially when you consider all employees as talents.

Potential

Another way of looking at talent is *potential*—the potential to take on higher management positions long-term. What defines potential?

To have potential, the person needs to have the abilities to develop management skills in order to take on a management position. The significant dimensions from the competence lemon needed to be a potential manager are personal capability, social capability, the ability to learn, and the ability to manage complexity.

Personal capability in this case says that you have the potential to develop self-confidence in different contexts and adapt your personal capabilities depending on the situation. Furthermore, having social capability means that you can interact with people on different levels and are able to build networks.

Moreover, you have the ability to learn in order to acquire new competence quickly. In addition, you have the ability to manage complexity, meaning managing difficult stakeholder situations, managing work in different contexts, and making decisions without having all facts. The development plan for a potential manager definitely includes the ability to act in different contexts.

The question is why leadership qualities, knowledge, and experience are less prioritized. Leadership qualities can be developed if you are strong in the four dimensions of the competence lemon mentioned above; those four dimensions are the basis to be a good leader, or at least to have good leadership qualities. Having good leadership qualities and being a good leader are two different things. The last dimension, knowledge and experience, is the easiest dimension to develop. As human beings, we can easily take in new knowledge, but changing behavior takes time and effort.

Competence Development

Another HR perspective is *competence development.* Using the competence lemon as a framework for competence development was discussed in an earlier section of this chapter, and HR's role in this process is to ensure that managers understand how the competence lemon supports competence development in different situations. Moreover, by understanding the different mechanisms in the competence loop, HR can support organizational leaders in how to interpret what new competences people have developed through working on projects or other activities.

6.8.2 Recruiting Talented Employees

Recruitment is another area in which HR needs to take a leading position. In many cases, line managers focus on a candidate's previous knowledge and experience when recruiting new employees. Ninety-five percent of a normal résumé or job application consists of previous knowledge and experience; the majority of the screening and the selection are based on this information, along with direct evaluation of how positions have been managed in the past rather than how the position will be managed in the future. Maybe some personal and social capabilities can be interpreted by reading between the lines in the résumé and the application.

In many cases, the interviews also focus on previous knowledge and experience but, in addition, show the applicant's social capability. It is possible that, by narrowly focusing on previous knowledge and experience, the interviewer could place too much of the decision on the social capability, possibly recruiting a candidate who is great at selling themselves but may be a low performer. If the applicant has a high degree of social capability and the interviewer focuses too much on knowledge and experience, not taking in this dimension of competence, the applicant can use his or her social capability to oversell their knowledge and experience, which could lead to recruiting people that have neither the right competence nor the right potential.

Therefore, instead of focusing on previous knowledge and experience, recruitment should focus on the other dimensions of competence from the competence lemon and on how the applicant fits into the context and the organizational culture. In addition, recruitment should focus on the future and not on the past, meaning that the position will change and the candidate must be able to grow with the position and be able to change their way of work to adapt to changing market conditions. As mentioned before, knowledge is the dimension of competence that is easiest to acquire, whereas our behavior does not change easily.

Other HR Concerns

Because both the competence loop and the competence lemon rely on organizational culture and identity, HR can work with organizational values and behaviors, considering all aspects of organizational culture and identity—namely, corporate, national, industry, and professional—with the goal of establishing a strong and functional organizational culture. In this context, a functional culture means one that is an asset to the organization and whose values are constructive and support the organizational goals (Flamholtz and Randle 2011).

Some perspectives should be considered from HR's point of view in terms of effective competence management. HR can take an active role, being the engine in many of the processes with respect to managing competence, in both social processes—factors such as generating new competence and establishing a functional culture—as well as in organizational processes such as talent management and recruitment. In addition, HR should support the organization in the use of tools, such as REPI.

The next section will move on to what organizations should do and what they should not do, or what they should start to do or stop doing.

6.9 Dos and Don'ts

This section will bring together things that I consider that organizations should and should not do, or maybe what organizations should start doing or stop doing, to be effective in competence management—in other words, the dos and don'ts. What is right and wrong? One of the main messages in this book is that everything depends on the context in which we act. How can I then say what to do and what not to do? The answer is simple: I cannot, but I can *recommend* what to do and not to do. My belief is that the dos and don'ts are valid for almost every knowledge-intensive organization relying on internal and external knowledge workers. As human beings focused on problems, it is always easier to propose the don'ts, but I will try balance these with some dos.

6.9.1 Dos and Don'ts of Recruitment

- **Do** consider all dimensions of the competence loop when recruiting new employees. Too much focus on knowledge and experience alone leads to poor recruitments. The résumé is, in general, a description of a person's past life, not of what the person can achieve in the future.
- **Don't** recruit only to the current role, or what the role looked like in the past; instead recruit people for a future role based on their potential. Too much unsuccessful recruitment has been done looking at the one to replace instead of recruiting for the future.

6.9.2 Dos and Don'ts of Competence Development

- **Do** build competence development into daily work. See daily work, and especially projects, as learning arenas in which people learn new things and gain new competence. Work with the competence goals that should be achieved in daily work. The competence lemon framework and REPI are excellent support here. Believe that competence evolves by trying new and unknown things, solving problems, reflecting on what is going well and what is going badly, knowledge sharing, and other factors that are described in this book.
- **Don't** consider courses as equivalent to competence development. Courses are good, but they are generally only triggers to start gaining new knowledge.

6.9.3 Dos and Don'ts of Performance Management

Performance management is another area in which traditional appraisal processes should be avoided for knowledge workers. The world is moving too fast to set goals once a year and follow them up.

- **Do** start to have a more agile approach to performance management, with continuous talks between the employee and the line manager. Not only are REPI meetings an excellent way of following performance, they are also a good competence development activity. I call for agile performance management with REPI as a base, and continuous communication between employee and line manager as well as among the people working in the organization. This too is a knowledge-sharing activity that will contribute to organizational learning.
- **Do** work on establishing a positive and functional organizational culture, considering corporate culture, national culture, industry culture, and professional culture. The organizational culture should be based on the 3Ts: trust, transparency, and togetherness. If people feel trust in the organization, they will share knowledge, dare to try new things, and give feedback in a positive way. In an organization with transparency, people will trust information, create positive energy, and do the rights things. When people feel togetherness, they collaborate to a greater extent, work toward the same goals, and learn from each other. Moreover, if the organizational culture is weak, it is possible to establish a subculture in a project by working with communication and norms.

6.9.4 Dos and Don'ts of Context

Another area to consider is the context in which a project, or any other task, will be performed.

- **Don't** believe that a person who excels in one context will automatically be a high performer in another. Context has a major impact on the application of competence.
- **Do** consider the context when selecting a person for a role. The context can be the company, the market, the role, the culture, and much more.
- **Do** also consider the context when prioritizing which factors of the competence loop should be the focus. An organization heavily dependent on external competence needs to focus on identifying internal/external competences, absorbing capacity, external sourcing, etc. Another organization might be immature in project management and needs to focus on developing a working project governance structure.

There are many dos and don'ts to consider when managing project competence. Those above are just a few, but ones that I consider are crucial for effective competence management.

6.9.5 The Final Do: Reflection

The last do I would like to recommend is *to reflect*. To reflect is perhaps individually the most important factor in gaining new competence. Reflection is the first letter in REPI and is the way to analyze situations, problems, information, people's actions, and everything else that we encounter in daily life. People can reflect on their own or in a group. If people are reflecting in a group, the knowledge-sharing activity will have an even better result. Projects and other teams should take time to reflect together: this can be done as lessons learned in projects, but do it often and not just at the end of the project, and include what new competences people have gained in the project.

The agile project methodology Scrum has a retrospective meeting after every sprint, focusing on how the process worked during the last sprint. The retrospective meeting could also include what new competences people have gained during the last sprint. Other types of teams should also build in reflection in groups as a way of working, including in this activity what people have learned.

Individual reflection is not simply competence generating: it is also good for your body and life–work balance. Using some time during the day to reflect on what you are doing, what is going well and what is going less well, will give you new insights into yourself and what new learning you have had that particular day. We learn every hour and every day, and we need to take the opportunity to take in these facts and understand that competence is not static: it is continuously evolving.

Glossary

Term	Definition
Ability to learn	The ability to continuously acquire new competence and use it in a productive way.
Ability to manage complexity	The ability to manage ambiguity and complex, constantly changing situations, such as complex stakeholder situations, many different suppliers, etc. It also means the ability to take in information and link different domains to a conclusion.
Absorptive capacity	The ability to absorb external information and knowledge, integrate it with internal knowledge, and apply it in a way that contributes to the organizational goals. Learning from customers, suppliers, and other external parties in order to achieve competitive advantage.
Adaptive capabilities	An organization's ability to adapt in a fast-moving market through means of strategic flexibility and balancing its exploration and exploitation strategies.
Agile performance management	The way leaders and managers follow up employees' performance continuously, in contrast to yearly performance appraisal.
Capabilities	An organization's capacity to deploy resources, encapsulating both explicit processes and tacit elements, such as know-how and leadership, meaning that capabilities are embedded in processes.
Competence management	The way for an organization to manage its competences at the corporation, group, and individual levels in order to be innovative and achieve a competitive advantage.

Term	Definition
Core competences	The capabilities that are critical to a business in achieving a competitive advantage.
Dynamic capabilities	The ability to achieve new forms of competitive advantage—that is, how organizations can demonstrate timely responsiveness and respond to the market's need for innovation in a rapid and flexible manner, combined with the management capability to coordinate and redeploy external and internal competences efficiently.
Explicit knowledge	Coded and stored knowledge built into working processes, documentation, information, etc.
Exploitative learning	Consists of two parts: (1) transforming new competences into the organizational competence base, and (2) utilizing competences in new projects or other value-creation activities.
Exploratory learning	Consists of two parts: (1) the part in which people develop new competences, and (2) the part in which the new competences are interpreted and understood by others in the organization.
HRM competence management practices	A subset of human resource management practices connected to managing competence.
Incremental innovation	Reinforcement and exploitation of existing knowledge in order to improve current products and services.
Individual learning	Activities that develop all dimensions of competence.
Innovative capabilities	The ability to align strategic innovative orientation with innovative processes and behavior, thereby developing new products, services, or markets.
Knowledge intensity	A context in which new knowledge is acquired through problem solving, experimentation, or learning.
Knowledge-intensive organization	An organization in which knowledge has more importance than other inputs and human capital dominates. Organizations have different levels of knowledge-intensity, e.g. a research and development organization is more knowledge-intensive than an organization producing cars.
Leadership qualities	The qualities to provide information to others, enabling them to solve a problem. Examples are people management skills, having the capability to lead a team in the right direction, and the ability to lift and support others in their work.

Term	Definition
Learning	A permanent change in behavior based on experience and reception, which leads to better performance.
Learning capabilities	Capabilities that enable the generation of new competence. A combination of practices that promotes organizational and individual learning and an open culture that promotes sharing of knowledge.
Organizational culture	How a group shares values, beliefs, goals, and expectations that will persist over time, even when group members have changed. Organizational culture will in this book have the following perspectives: corporate, national, profession, and industry.
Organizational identity	How the organization allows itself to be known through a set of meanings, which also allows people to describe and remember it. The identity of an organization can also be described as what it expects to be and what it stands for.
Organizational learning	A process of acquiring, transferring, and integrating new knowledge and, in this way, adding value to the organization. Organizational learning can be carried out by both informal and formal processes. The informal processes occur when people share knowledge in daily work, and the formal processes are the means by which the organization integrates knowledge from individual to group to organizational levels, and in this way either expands existing knowledge or creates new knowledge.
PBO	Project-based organization. An organization that carries out most of its activities in the form of projects, and the project dimension is stronger than the functional dimension. In many cases, a PBO does not have any functional organization at all—for instance, a conference or a specific sporting event.
Personal capabilities	The combination of personal characteristics (e.g., being pedagogic or innovative) and a person's attitude to work (e.g., responsible or positive).
Project-intensive organization	An organization in which there is a coexistence of a functional organization and projects, and where a considerable part of the organization's activities is conducted through projects.
Radical innovation	Exploration of new knowledge in order to develop new products, services, or markets.
Knowledge workers	People who work in knowledge-intensive organizations need to acquire new knowledge to perform their jobs in a good way

Term	Definition
Social capabilities	A person's ability to share knowledge and interact with others by listening and being open to others' ideas and opinions, combined with the person's ability to explain your knowledge to others.
Subcultures	A culture in a specific part of the organization—for instance, a project with specific norms, behaviors, and paths to communicate.
Sustainable competence	Also called dynamic competence. Competence that is renewed to adapt to new conditions and ways of working. New knowledge is acquired based on need and in which context the competence shall be applied.
Tacit knowledge	Knowledge embedded in practices and what we have in our subconscious minds. This kind of knowledge is not codified or written down but exists only in our minds.

Bibliography

Agha, S., Alrubaiee, L., & Jamhour, M. (2011). Effect of Core Competence on Competitive Advantage and Organizational Performance. *International Journal of Business and Management, 7*(1), 192–204.

Ajmal, M. M., & Koskinen, K. U. (2008). Knowledge Transfer in Project-Based Organizations: An Organizational Culture Perspective. *Project Management Journal, 39*(1), 7–15.

Akbar, H., & Mandurah, S. (2014). Project-Conceptualisation in Technological Innovations: A Knowledge-Based Perspective. *International Journal of Project Management, 32*(5), 759–772.

Alvesson, M. (2000). Social Indentity and the Problem of Loyalty in Knowledge-Intensive Companies. *Journal of Management Studies, 37*(8), 1101–1124.

Alvesson, M. (2011). De-Essentializing the Knowledge Intensive Firm: Reflections on Sceptical Research Going Against the Mainstream. *Journal of Management Studies, 48*(7), 1640–1661.

Argote, L., & Ingram, P. (2000). Knowledge Transfer: A Basis for Competitive Advantage in Firms. *Organizational Behavior and Human Decision Processes, 82*(1), 150–169.

Artto, K., & Kujala, J. (2008). Project Business as a Research Field. *International Journal of Managing Projects in Business, 1*(4), 469–497.

Arvidsson, N. (2009). Exploring Tensions in Projectified Matrix Organisations. *Scandinavian Journal of Management, 25*(1), 97–107.

Barr, R. B., & Tagg, J. (1995). From Teaching to Learning: A New Paradigm for Undergraduate Education. *Change 27*(6), 12–25.

Belbin, R. M. (2012). *Team Roles at Work*. London: Routledge.

Biedenbach, T., & Müller, R. (2012). Absorptive, Innovative and Adaptive Capabilities and Their Impact on Project and Project Portfolio Performance. *International Journal of Project Management, 30*(5), 621–635.

Blindenbach-Driessen, F., & Van Den Ende, J. (2010). Innovation Management Practices Compared: The Example of Project-Based Firms. *Journal of Product Innovation Management, 27*(5), 705–724.

Bloom, B. S. (1956). *Taxonomy of Educational Objectives: The Classification of Educational Goals. Handbook 1, Cognitive Domain.* New York: David McKay.

Bono, E. D. (1994). *Parallel Thinking: From Socratic Thinking to de Bono Thinking.* London: Penguin Group.

Brady, T., & Davies, A. (2004). Building Project Capabilities: From Exploratory to Exploitative Learning. *Organization Studies, 25*(9), 1601–1621.

Bredin, K. (2008). People Capability of Project-Based Organisations: A Conceptual Framework. *International Journal of Project Management, 26*(5), 566–576.

Bresman, H., & Zellmer-Bruhn, M. (2013). The Structural Context of Team Learning: Effects of Organizational and Team Structure on Internal and External Learning. *Organization Science, 24*(4), 1120–1139.

Cabello-Medina, C., López-Cabrales, Á., & Valle-Cabrera, R. (2011). Leveraging the Innovative Performance of Human Capital Through HRM and Social Capital in Spanish Firms. *The International Journal of Human Resource Management, 22*(4), 807–828.

Cepeda, G., & Vera, D. (2007). Dynamic Capabilities and Operational Capabilities: A Knowledge Management Perspective. *Journal of Business Research, 60*(5), 426–437.

Chen, H., & Chang, W. (2011). Core Competence: What "Core" You Mean? From a Strategic Human Resource Management Perspective. *African Journal of Business Management, 5*(14), 5738–5745.

Clardy, A. (2008). Human Resource Development and the Resource-Based Model of Core Competencies: Methods for Diagnosis and Assessment. *Human Resource Development Review, 7*(4), 387–407.

Crawford, L. (2005). Senior Management Perceptions of Project Management Competence. *International Journal of Project Management, 23*(1), 7–16.

Dale, E. (1969). *Audiovisual Methods in Teaching* (3rd ed.). 1969.

De Wever, S. (2008). Learning and Capability Development: The Impact of Social Capital. In Heene, A., Martens, R., & Sanchez, R. (Eds.), *Advances in Applied Business Strategy*, Vol. 10, pp. 121–157: Emerald Group Publishing Limited.

Eden, C., & Ackermann, F. (2010). Competences, Distinctive Competences, and Core Competences. *Research in Competence-Based Management, 5*, 3–33.

Eisenhardt, K. M., & Martin, J. A. (2000). Dynamic Capabilities: What Are They? *Strategic Management Journal, 21*(10–11), 1105–1121.

Eriksson, P. E. (2013). Exploration and Exploitation in Project-Based Organizations: Development And Diffusion of Knowledge at Different Organizational Levels in Construction Companies. *International Journal of Project Management, 31*(3), 333–341.

Eriksson, T. (2014). Processes, Antecedents and Outcomes of Dynamic Capabilities. *Scandinavian Journal of Management, 30*(1), 65–82.

Fenwick, T. (2008). Understanding Relations of Individual–Collective Learning in Work: A Review of Research. *Management Learning, 39*(3), 227–243.

Flamholtz, E., & Randle, Y. (2011). *Corporate Culture: The Ultimate Strategic Asset.* Palo Alto: Stanford University Press.

Galbraith, J. R. (1971). Matrix Organization Designs: How to Combine Functional and Project Forms. *Business Horizons, 14*(1), 29–40.

Garri, M., Konstantopoulos, N., & Bekiaris, M. (2013). Corporate Strategy, Corporate Culture & Customer Information. *Procedia – Social and Behavioral Sciences, 73*, 669–677.

Gilbert, M., & Cordey-Hayes, M. (1996). Understanding the Process of Knowledge Transfer to Achieve Successful Technological Innovation. *Technovation, 16*(6), 301–312.

Guiso, L., Sapienza, P., & Zingales, L. (2008). Alfred Marshall Lecture: Social Capital as Good Culture. *Journal of the European Economic Association, 6*(2–3), 295–320.

Gunsel, A., Siachou, E., & Acar, A. Z. (2011). Knowledge Management and Learning Capability to Enhance Organizational Innovativeness. *Procedia – Social and Behavioral Sciences, 24*(0), 880–888.

Gupta, S., Woodside, A., Dubelaar, C., & Bradmore, D. (2009). Diffusing Knowledge-Based Core Competencies for Leveraging Innovation Strategies: Modelling Outsourcing to Knowledge Process Organizations (KPOs) in Pharmaceutical Networks. *Industrial Marketing Management, 38*(2), 219–227.

Hager, P., & Gonczi, A. (1996). What Is Competence? *Medical Teacher, 18*(1), 15–18.

Hamel, G., & Prahalad, C. K. (1994). Competing for the Future; What Drives Your Company's Agenda: Your Competitor's View of the Future or Your Own? (Adapted from Hamel, G., & Prahalad, C. K., *Competing for the Future.*) *Harvard Business Review, 72*(4), 122.

Harzallah, M., Berio, G., & Vernadat, F. (2006). Analysis and Modeling of Individual Competencies: Toward Better Management of Human Resources. *IEEE Transactions on Systems, Man and Cybernetics, Part A: Systems and Humans, 36*(1), 187–207.

Hatch, M. J., & Cunliffe, A. L. (2006). *Organization Theory: Modern, Symbolic, and Postmodern Perspectives*. Oxford: Oxford University Press.

Hobday, M. (2000). The Project-Based Organisation: An Ideal Form for Managing Complex Products and Systems? *Research Policy, 29*(7–8), 871–893.

Hofstede, G. (1998). Identifying Organizational Subcultures: An Empirical Approach. *Journal of Management Studies, 35*(1), 1–12.

Holmqvist, M. (2003). A Dynamic Model of Intra- and Interorganizational Learning. *Organization Studies, 24*(1), 95–123.

Hubbard, G., Zubac, A., & Johnson, L. (2008). Linking Learning, Customer Value, and Resource Investment Decisions: Developing Dynamic Capabilities. *Advances in Applied Business Strategy, 10*, 9–27.

Hung, I. W., Choi, A. C. K., & Chan, J. S. F. (2003). An Integrated Problem-Based Learning Model for Engineering Education. *International Journal of Engineering Education, 19*(5), 734–737.

Jasimuddin, S. M. (2014). Face-to-Face Interface in Software Development: Empirical Evidence from a Geographically Dispersed High-Tech Laboratory. *International Journal of Technology and Human Interaction, 10*(1), 48–60.

Jerez-Gómez, P., Céspedes-Lorente, J., & Valle-Cabrera, R. (2005). Organizational Learning Capability: A Proposal of Measurement. *Journal of Business Research, 58*(6), 715–725.

Jiménez-Barrionuevo, M. M., García-Morales, V. J., & Molina, L. M. (2011). Validation of an Instrument to Measure Absorptive Capacity. *Technovation, 31*(5–6), 190–202.

Jiménez-Jiménez, D., & Sanz-Valle, R. (2011). Innovation, Organizational Learning, and Performance. *Journal of Business Research, 64*(4), 408–417.

Kang, J., Rhee, M., & Kang, K. H. (2010). Revisiting Knowledge Transfer: Effects of Knowledge Characteristics on Organizational Effort for Knowledge Transfer. *Expert Systems with Applications, 37*(12), 8155–8160.

Kang, S.C., Morris, S. S., & Snell, S. A. (2007). Relational Archetypes, Organizational Learning, and Value Creation: Extending the Human Resource Architecture. *Academy of Management Review, 32*(1), 236–256.

Kärreman, D. (2010). The Power of Knowledge: Learning from "Learning by Knowledge-Intensive Firm." *Journal of Management Studies, 47*(7), 1405–1416.

Keegan, A., Huemann, M., & Turner, J. R. (2012). Beyond the Line: Exploring the HRM Responsibilities of Line Managers, Project Managers and the HRM Department in Four Project-Oriented Companies in the Netherlands, Austria, the UK and the USA. *The International Journal of Human Resource Management, 23*(15), 3085–3104.

Killen, C. P., Hunt, R. A., & Kleinschmidt, E. J. (2008). Learning Investments and Organizational Capabilities; Case Studies on the Development of Project Portfolio Management Capabilities. *International Journal of Managing Projects in Business, 1*(3), 334–351.

Kim, A., & Lee, C. (2012). How Does HRM Enhance Strategic Capabilities? Evidence from the Korean Management Consulting Industry. *International Journal of Human Resource Management, 23*(1), 126–146.

Kocoglu, I., Imamoglu, S. Z., Ince, H., & Keskin, H. (2012). Learning, R&D and Manufacturing Capabilities as Determinants of Technological Learning: Enhancing Innovation and Firm Performance. *Procedia – Social and Behavioral Sciences, 58*(0), 842–852.

Koskinen, K. U. (2009). Project-Based Company's Vital Condition: Structural Coupling. An Autopoietic View. *Knowledge and Process Management, 16*(1), 13–22.

Koskinen, K. U. (2015). Processual Knowledge Production in Organisations Dealing with Projects. *International Journal of Project Organisation and Management, 7*(3), 206–220.

Kotter, J. P., & Heskett, J. L. (1992). *Corporate Culture and Performance.* New York: The Free Press.

Krathwohl, D. R. (2002). A Revision of Bloom's Taxonomy: An Overview. *Theory into Practice, 41*(4), 212–218.

Laakso-Manninen, R., & Viitala, R. (2007). *Competence Management and Human Resource Development. A Theoretical Framework for Understanding the Practices of Modern Finnish Organizations.* Helsinki: HAAGA-HELIA University of Applied Sciences.

Larson, E. W., & Gobeli, D. H. (1987). Matrix Management: Contradictions and Insights. *California Management Review, 29*(4), 126.

Le Deist, F. D., & Winterton, J. (2005). What Is Competence? *Human Resource Development International, 8*(1), 27–46.

Lengnick-Hall, M. L., Lengnick-Hall, C. A., Andrade, L. S., & Drake, B. (2009). Strategic Human Resource Management: The Evolution of the Field. *Human Resource Management Review, 19*(2), 64–85.

Li, C.Y. (2012). Knowledge Stickiness in the Buyer–Supplier Knowledge Transfer Process: The Moderating Effects of Learning Capability and Social Embeddedness. *Expert Systems with Applications, 39*(5), 5396–5408.

Liao, S., Chang, W., Hu, D., & Yueh, Y. (2012). Relationships among organizational culture, knowledge acquisition, organizational learning, and organizational innovation in Taiwan's banking and insurance industries. *The International Journal of Human Resource Management, 23*(1), 52-70.

Liao, S.H., Chang, W.J., & Wu, C.C. (2010). An Integrated Model for Learning Organization with Strategic View: Benchmarking in the Knowledge-Intensive Industry. *Expert Systems with Applications, 37*(5), 3792–3798.

Lin, H. E., McDonough, E. F., Lin, S. J., & Lin, C. Y. Y. (2013). Managing the Exploitation/Exploration Paradox: The Role of a Learning Capability and Innovation Ambidexterity. *Journal of Product Innovation Management, 30*(2), 262–278.

Lindkvist, L. (2004). Governing Project-Based Firms: Promoting Market-Like Processes Within Hierarchies. *Journal of Management and Governance, 8*(1), 3–25.

Lisboa, A., Skarmeas, D., & Lages, C. (2011). Innovative Capabilities: Their Drivers and Effects on Current and Future Performance. *Journal of Business Research, 64*(11), 1157–1161.

Lopez-Cabrales, A., Real, J. C., & Valle, R. (2011). Relationships Between Human Resource Management Practices and Organizational Learning Capability: The Mediating Role of Human Capital. *Personnel Review, 40*(3), 344–363.

Male, S. A., Bush, M. B., & Chapman, E. S. (2011). An Australian Study of Generic Competencies Required by Engineers. *European Journal of Engineering Education, 36*(2), 151–163.

March, J. G. (1991). Exploration and Exploitation in Organizational Learning. *Organization Science, 2*(1), 71–87.

Maurer, I., Bartsch, V., & Ebers, M. (2011). The Value of Intra-Organizational Social Capital: How It Fosters Knowledge Transfer, Innovation Performance, and Growth. *Organization Studies, 32*(2), 157–185.

Medina, A., Müller, R., & Bredillet, C. (2011). *The Unresolved Struggles Between Project Managers and Functional Managers in Matrix Organizations.* Paper presented at the Proceedings of the 11th Annual European Academy of Management, Estonian Business School, Tallinn, Estonia.

Medina, R., & Medina, A. (2014). The Project Manager and the Organisation's Long-Term Competence Goal. *International Journal of Project Management, 32*(8), 1459–1470.

Melewar, T. C. (2003). Determinants of the Corporate Identity Construct: A Review of the Literature. *Journal of Marketing Communications, 9*(4), 195–220.

Melkonian, T., & Picq, T. (2011). Building Project Capabilities in PBOs: Lessons from the French Special Forces. *International Journal of Project Management, 29*(4), 455–467.

Nonaka, I. (1994). A Dynamic Theory of Organizational Knowledge Creation. *Organization Science, 5*(1), 14–37.

Oltra, V., & Vivas-López, S. (2013). Boosting Organizational Learning Through Team-Based Talent Management: What Is the Evidence from Large Spanish Firms? *The International Journal of Human Resource Management, 24*(9), 1853–1871.

Passow, H. J. (2012). Which ABET Competencies Do Engineering Graduates Find Most Important in Their Work? *Journal of Engineering Education, 101*(1), 95–118.

Pemsel, S., & Müller, R. (2012). The Governance of Knowledge in Project-Based Organizations. *International Journal of Project Management, 30*(8), 865–876.

Peteraf, M. A. (1993). The Cornerstones of Competitive Advantage: A Resource-Based View. *Strategic Management Journal, 14*(3), 179–191.

Pinto, J. K., & Rouhiainen, P. (2001). *Building Customer-Based Project Organizations.* Chichester: Wiley.

Polanyi, M. (1958). *Personal Knowledge: Towards a Post-Critical Philosophy.* Chicago: University of Chicago Press.

Popaitoon, S., & Siengthai, S. (2014). The Moderating Effect of Human Resource Management Practices on the Relationship Between Knowledge Absorptive Capacity and Project Performance in Project-Oriented Companies. *International Journal of Project Management, 32*(6), 908–920.

Prahalad, C. K., & Hamel, G. (1990). The Core Competence of the Corporation. *Harvard Business Review, 68*(3), 79–91.

Pratt, M. G., Rockmann, K. W., & Kaufmann, J. B. (2006). Constructing Professional Identity: The Role of Work and Identity Learning Cycles in the Customization of Identity Among Medical Residents. *The Academy of Management Journal, 49*(2), 235–262.

Project Management Institute. (2007). *Project Manager Competency Development Framework* (2nd ed.). Newtown Square, PA: Project Management Institute.

Reich, B. H., Gemino, A., & Sauer, C. (2008). Modeling the Knowledge Perspective of IT Projects. *Project Management Journal, 39*(S1).

Reich, B. H., Gemino, A., & Sauer, C. (2012). Knowledge Management and Project-Based Knowledge in IT Projects: A Model and Preliminary Empirical Results. *International Journal of Project Management, 30*(6), 663–674.

Reilly, P., & Williams, T. (2012). *Global HR: Challenges Facing the Function.* Farnham: Gower.

Rynes, S., Colbert, A., & Brown, K. (2002). HR Professionals' Beliefs About Effective Human Resource Practices: Correspondence Between Research and Practice. *Human Resource Management, 41*(2), 149–174.

Sanford, B. (1989). *Strategies for Maintaining Professional Competence: A Manual for Professional Associations and Faculties.* Toronto: Canadian Scholars' Press.

Schutz, W. (1958). *FIRO: A Three-Dimensional Theory of Interpersonal Behavior.* New York: Rinehart & Winston.

Shenhar, A. J. (2001). One Size Does Not Fit All Projects: Exploring Classical Contingency Domains. *Management Science, 47*(3), 394–414.

Shih, H.-A., & Chiang, Y.-H. (2003). Exploring Relationships Between Corporate Core Competence, Corporate Strategy, and HRM Practices in Training Institutions. *Asia Pacific Management Review, 8*(3), 281–310.

Sinha, K. K., & Van de Ven, A. H. (2005). Designing Work Within and Between Organizations. *Organization Science, 16*(4), 389–408.

Söderlund, J. (2005). *Projektledning & projektkompetens: Perspektiv på konkurrenskraft.* Malmö: Liber.

Söderlund, J., & Bredin, K. (2006). HRM in Project-Intensive Firms: Changes and Challenges. *Human Resource Management, 45*(2), 249–265.

Spencer, L. M., McClelland, D. C., & Kelner, S. (1994). *Competency Assessment Methods: History and State of the Art.* Boston: Hay/McBer.

Starbuck, W. (1992). Learning by Knowledge-Intensive Firms. *Journal of Management Studies, 29*(6), 713–740.

Subramaniam, M., & Youndt, M. A. (2005). The Influences of Intellectual Capital on the Types of Innovative Capabilities. *Academy of Management Journal, 48*(3), 450–463.

Swart, J., & Kinnie, N. (2003). Knowledge-Intensive Firms: The Influence of the Client on HR Systems. *Human Resource Management Journal, 13*(3), 37–55.

Szulanski, G. (2000). The Process of Knowledge Transfer: A Diachronic Analysis of Stickiness. *Organizational Behavior and Human Decision Processes, 82*(1), 9–27.

Tamayo-Torres, I., Ruiz-Moreno, A., & Verdú, A. J. (2010). The Moderating Effect of Innovative Capacity on the Relationship Between Real Options and Strategic Flexibility. *Industrial Marketing Management, 39*(7), 1120–1127.

Teece, D. J., Pisano, G., & Shuen, A. (1997). Dynamic Capabilities and Strategic Management. *Strategic Management Journal, 18*(7), 509–533.

Teodorescu, T. (2006). Competence versus Competency: What Is the Difference? *Performance Improvement, 45*(10), 27–30.

Turner, J. R. (1999). *The Handbook of Project-Based Management: Improving the Processes for Achieving Strategic Objectives.* London: McGraw-Hill.

Turner, J. R. (2014). *Gower Handbook of Project Management* (5th ed.). Surrey: Gower Publishing Limited.

Turner, J. R., & Müller, R. (2006). *Choosing Appropriate Project Managers: Matching Their Leadership Style to the Type of Project.* Newtown Square, PA: Project Management Institute.

Volberda, H. W., Foss, N. J., & Lyles, M. A. (2010). Perspective – Absorbing the Concept of Absorptive Capacity: How to Realize Its Potential in the Organization Field. *Organization Science, 21*(4), 931–951.

Von Krogh, G., & Roos, J. (1996). Five Claims on Knowing. *European Management Journal, 14*(4), 422–426.

Walker, D. H., & Lloyd-Walker, B. M. (2015). *Collaborative Project Procurement Arrangements.* Newtown Square, PA: Project Management Institute.

Wang, C. L., & Ahmed, P. K. (2007). Dynamic Capabilities: A Review and Research Agenda. *International Journal of Management Reviews, 9*(1), 31–51.

Whitley, R. (2006). Project-Based Firms: New Organizational Form or Variations on a Theme? *Industrial and Corporate Change, 15*(1), 77–99.

Winch, G. M. (2014). Three Domains of Project Organising. *International Journal of Project Management, 32*(5), 721–731.

Wright, P. M., Dunford, B. B., & Snell, S. A. (2001). Human Resources and the Resource Based View of the Firm. *Journal of Management, 27*(6), 701–721.

Zollo, M., & Winter, S. G. (2002). Deliberate Learning and the Evolution of Dynamic Capabilities. *Organization Science, 13*(3), 339–351.

Index

T